Aeronautical Chart User's Guide

9th Edition

Aeronautical Chart User's Guide

9th Edition

Federal Aviation Administration

National Aeronautical Navigation Services

SKYHORSE PUBLISHING

Skyhorse Publishing books may be purchased in bulk at special discounts for sales promotion, corporate gifts, fund-raising, or educational purposes. Special editions can also be created to specifications. For details, contact the Special Sales Department, Skyhorse Publishing, 307 West 36th Street, 11th Floor, New York, NY 10018 or info@skyhorsepublishing.com.

Skyhorse® and Skyhorse Publishing® are registered trademarks of Skyhorse Publishing, Inc.®, a Delaware corporation.

www.skyhorsepublishing.com

10 9 8 7 6 5 4 3 2 1

Library of Congress Cataloging-in-Publication Data is available on file.

ISBN 978-1-61608-534-6

Printed in China

Table of Contents

VFR AERONAUTICAL CHARTS

EXPLANATION OF VFR TERMS AND SYMBOLS

The discussions and examples in this section are based on the Sectional Aeronautical Chart (Sectionals). Sectionals include the most current data and are at a scale (1:500,000) most beneficial to pilots flying under Visual Flight Rules. A pilot should have little difficulty in reading these charts which are, in many respects, similar to automobile road maps. Each chart is named for a major city within its area of coverage.

The chart legend lists various aeronautical symbols as well as information concerning drainage, terrain and contour elevations. You may identify aeronautical, topographical, and obstruction symbols (such as radio and television towers) by referring to the legend. Many landmarks which can be easily recognized from the air, such as stadiums, pumping stations, refineries, etc., are identified by brief descriptions adjacent to small black squares marking their exact locations ▪ cabin . Oil wells are shown by small open circles ○ oil . Water, oil and gas tanks are shown by small black circles • water and labeled accordingly, if known. The scale of an item may be increased to make it easier to read on the chart.

AeroNav Services' charts are prepared in accordance with specifications of the Interagency Air Cartographic Committee (IACC) and are approved by representatives of the Federal Aviation Administration (FAA) and the Department of Defense (DoD).

HYDROGRAPHY

Two tones of blue are used to distinguish water areas identified as "Open Water" and "Inland Water".

Open Water is defined as the limits (shorelines) of all coastal features at mean high water for oceans, seas and associated waters such as bays, gulfs, sounds, fords, large estuaries, etc. Exceptionally large lakes such as the Great Lakes, Great Bear Lake, Great Slave Lake, etc., will be considered as Open Water features. The Open Water tone will be extended inland as far as deemed necessary to adjoin the Inland Water tone (generally where drainage lines coalesce to a width of 0.1" approximate).

Inland Water is defined as all other bodies of water. Cartographic judgement is used as required in some instances.

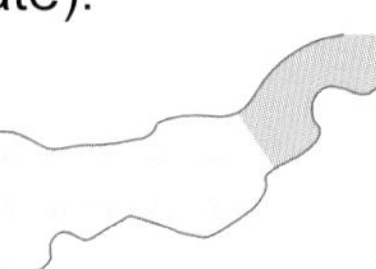

TERRAIN AND OBSTRUCTIONS

The elevation and configuration of the Earth's surface are certainly of prime importance to pilots. Cartographers devote a great deal of attention to showing relief and obstruction data in a clear and concise manner. Five different techniques are used: contour lines, shaded relief, color tints, obstruction symbols, and Maximum Elevation Figures (MEF).

1. Contour lines are lines connecting points on the Earth of equal elevation. On Sectionals, basic contours are spaced at 500' intervals. Intermediate contours may also be shown at 250' intervals in moderately level or gently rolling areas. Occasionally, auxiliary contours at 50, 100, 125, or 150' intervals may be used to portray smaller relief features in areas of relatively low relief. The pattern of these lines and their spacing gives the pilot a visual concept of the terrain. Widely spaced contours represent gentle slopes, while closely spaced contours represent steep slopes.

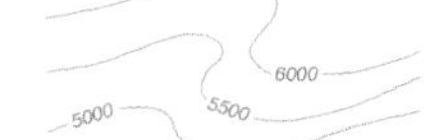

2. Shaded relief is a depiction of how the terrain might appear from the air. The cartographer shades the areas that would appear in shadow if illuminated by a light from the northwest. Studies have indicated that our visual perception has been conditioned to this view.

3. Color tints, also referred to as hypsometric tints, are used to depict bands of elevation relative to sea level. These colors range from light green for the lowest elevations to dark brown for the higher elevations.

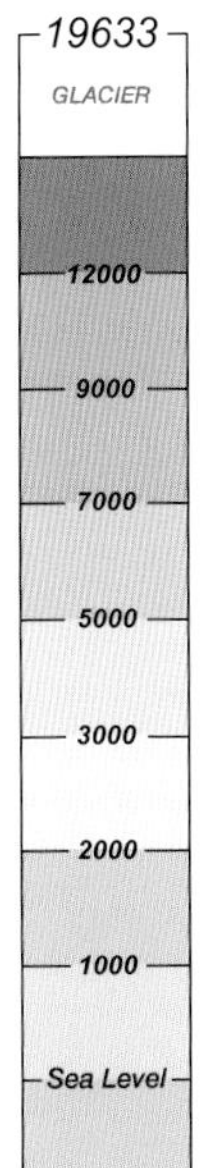

4. Obstruction symbols are used to depict man-made vertical features that may affect the National Airspace System. AeroNav Services maintains a database of nearly 127,000 obstacles in the United States, Canada, the Caribbean and Mexico. Each obstacle is evaluated by cartographers based on charting specifications before it is added to visual charts. When the position or elevation of an obstacle is unverified, it is marked UC (under construction or reported but not verified).

The data in the Digital Obstacle File (DOF) is collected and disseminated as part of Aero-

Nav Services' responsibility for depicting the National Airspace System.

Source data on terrain and obstructions is sometimes not complete or accurate enough for use in aeronautical publications; for example, a reported obstruction may be submitted with insufficient detail for determining the obstruction's position and elevation. Such cases are identified by AeroNav Services and investigated by the FAA Flight Edit program.

The FAA Flight Edit crew conducts data verification missions, visually verifying cultural and topographic features and reviewing all obstacle data. Charts are generally flight-checked every three years. This review includes checking for obstructions that may have been constructed, altered, or dismantled without proper notification.

Generally, only man-made structures extending more than 200' above ground level (AGL) are charted on Sectionals and TACs except within yellow city tint. Objects 200' or less are charted only if they are considered hazardous obstructions; for example, an obstruction is much higher than the surrounding terrain or very near an airport. Examples of features considered hazardous obstacles to low level flight are smokestacks, tanks, factories, lookout towers, and antennas. On World Aeronautical Charts (WACs) only obstacles 500' AGL and higher are charted.

Obstacles less than 1000' AGL are shown by the symbol . Obstacles 1000' AGL and higher are shown by the symbol . Man-made features which are used by FAA Air Traffic Control as checkpoints may be represented with pictorial symbols shown in black with the required elevation data in blue.

The elevation of the top of the obstacle above mean sea level (MSL) and the height of the structure AGL are shown when known or when they can be reliably determined by the cartographer. The AGL height is shown in parentheses below the MSL elevation. In extremely congested areas the AGL values may be omitted to avoid confusion.

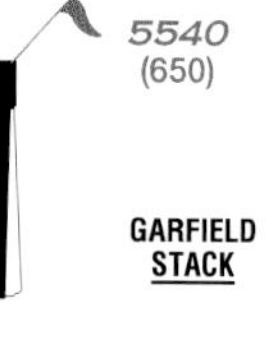

Obstacles are portrayed wherever possible. Since legibility would be impaired if all obstacles within city complexes or within high density groups of obstacles were portrayed, only the highest obstacle in an area is shown using 4977 (1432), the group obstacle symbol.

Obstacles under construction are indicated by the letters UC nearest to the obstacle type. If space is available, the AGL height of the obstruction is shown in parentheses; for example, (1501). Obstacles with high-intensity strobe lighting systems may operate part-time or by proximity activation and are shown as:

5. The Maximum Elevation Figure (MEF) represents the highest elevation, including terrain and other vertical obstacles (towers, trees, etc.), within a quadrant. A quadrant on Sectionals is the area bounded by ticked lines dividing each 30 minutes of latitude and each 30 minutes of longitude. MEF figures are depicted to the nearest 100' value. The last two digits of the number are not shown. In this example the MEF represents 12,500': 12 5. MEFs are shown over land masses as well as over open water areas containing man-made obstacles such as oil rigs.

In the determination of MEFs, extreme care is exercised to calculate the values based on the existing elevation data shown on source material. Cartographers use the following procedure to calculate MEFs:

When a man-made obstacle is more than 200' above the highest terrain within the quadrant:

1. Determine the elevation of the top of the obstacle above MSL.
2. Add the possible vertical error of the source material to the above figure (100' or 1/2 contour interval when interval on source exceeds 200'. U.S. Geological Survey Quadrangle Maps with contour intervals as small as 10' are normally used).
3. Round the resultant figure up to the next higher hundred foot level.

Example: Elevation of obstacle top (MSL) =		**2424**
Possible vertical error		**+100**
	equals	**2524**
Raise to the following 100' level		**2600**
Maximum Elevation Figure	**26**	

When a natural terrain feature or natural vertical obstacle (e.g. a tree) is the highest feature within the quadrangle:

1. Determine the elevation of the feature.
2. Add the possible vertical error of the source to the above figure (100' or 1/2 the contour interval when interval on source exceeds 200').
3. Add a 200' allowance for natural or man made obstacles which are not portrayed because they are below the minimum height at which the chart specifications require their portrayal.

4. Round the figure up to the next higher hundred foot level.

Example: Elevation of obstacle top (MSL) =	**3450**
Possible vertical error	**+100**
Obstacle Allowance	**+200**
equals	**3750**
Raise to the following 100' level	**3800**
Maximum Elevation Figure	**38**

Pilots should be aware that while the MEF is based on the best information available to the cartographer, the figures are not verified by field surveys. Also, users should consult the Aeronautical Chart Bulletin in the A/FD or AeroNav Services website to ensure that your chart has the latest MEF data available.

RADIO AIDS TO NAVIGATION

On visual charts, information about radio aids to navigation is boxed, as illustrated. Duplication of data is avoided. When two or more radio aids in a general area have the same name with different frequencies, TACAN channel numbers, or identification letters, and no misinterpretation can result, the name of the radio aid may be indicated only once within the identification box. VHF/UHF radio aids to navigation names and identification boxes (shown in blue) take precedence. Only those items that differ (e.g., frequency, Morse Code) are repeated in the box in the appropriate color. The choice of separate or combined boxes is made in each case on the basis of economy of space and clear identification of the radio aids.

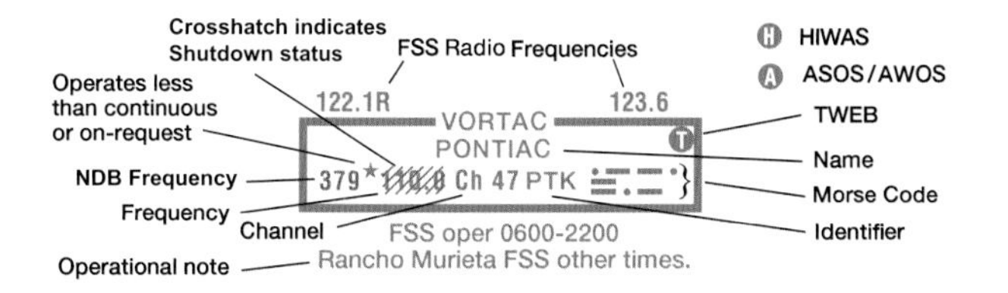

Radio aids to navigation located on an airport depicted by the pattern symbol may not always be shown by the appropriate symbol. A small open circle indicates the NAVAID location when colocated with an airport symbol. The type of radio aid to navigation will be identified by: VORTAC, VOR or VOR-DME, positioned on and breaking the top line of the navaid box.

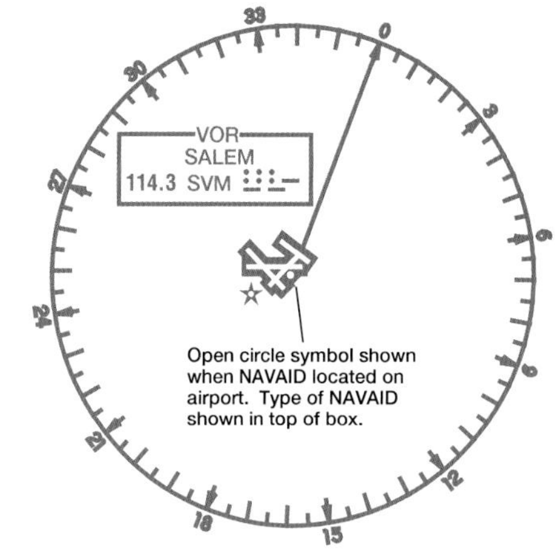

AIRPORTS

Airports in the following categories are charted as indicated (additional symbols are shown later in this Section).

Public use airports:

- Hard-surfaced runways greater than 8069' or some multiple runways less than 8069'
- Hard-surfaced runways 1500' to 8069'
- Other than hard-surfaced runways
- Seaplane bases

Military airports:

- Other than hard-surfaced runways

Hard-surfaced runways are depicted the same as public-use airports.

U.S. military airports are identified by abbreviations such as AAF (Army Air Field), AFB (Air Force Base), MCAS (Marine Corps Air Station), NAS (Naval Air Station), NAF (Naval Air Facility), NAAS Naval Auxiliary Air Station), etc. Canadian military airports are identified by the abbreviation DND (Department of National Defense).

Services available:

Tick marks around the basic airport symbol indicate that fuel is available and the airport is tended during normal working hours. (Normal working hours are Monday through Friday 10:00 A.M. to 4:00 P.M. local time.)

Other airports with or without services:

Airports are plotted in their true geographic position unless the symbol conflicts with a radio aid to navigation (navaid) at the same location. In such cases, the airport symbol will be displaced, but the relationship between the airport and the navaid will be retained.

Airports are identified by their designated name. Generic parts of long airport names (such as "airport", "field" or "municipal") and the first names of persons are commonly omitted unless they are needed to distinguish one airport from another with a similar name.

The figure at right illustrates the coded data that is provided along with the airport name. The elevation of an airport is the highest point on the usable portion of the landing areas. Runway length is the length of the longest active runway including displaced thresholds and excluding overruns. Runway length is shown to the nearest 100', using 70 as the division point; a runway 8070' in length is charted as 81, while a runway 8069' in length is charted as 80.

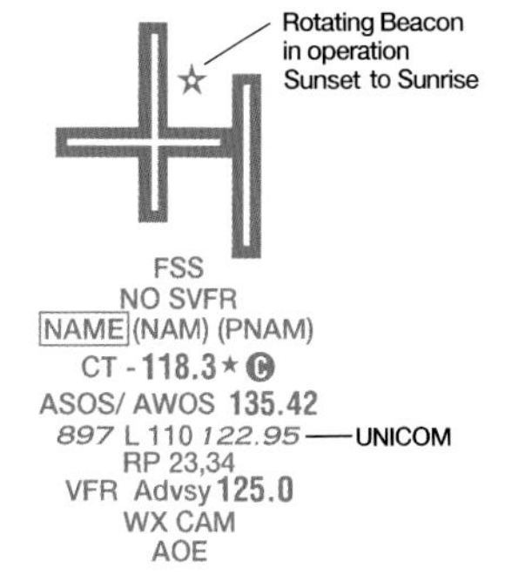

Airports with Control Towers (CT), and their related information, are shown in blue. All other airports, and their related information, are shown in magenta.

FSS - Flight Service Station on field
NO SVFR - Airports where fixed wing special visual flight rules operations are prohibited (shown above airport name) F.A.R. 91
[ ] - Indicates F.A.R. 93 Special Air Traffic Rules and Airport Traffic Patterns
(NAM) - Location Identifier
(PNAM) - ICAO Location Indicator
CT - 118.3 - Control Tower (CT) - primary frequency
★ - Star indicates operation part-time. See tower frequencies tabulation for hours of operation
C - Indicates Common Traffic Advisory Frequencies (CTAF) (Not shown on WAC)
ATIS 123.8 - Automatic Terminal Information Service
ASOS/ AWOS 135.42 - Automated Surface Weather Observing Systems (shown when full-time ATIS is not available.) Some ASOS/AWOS facilities may not be located at airport. (Not shown on WAC)
897 - Elevation in feet
L - Lighting in operation Sunset to Sunrise
*L - Lighting limitations exist; refer to Airport/Facility Directory.
110 - Length of longest runway in hundreds of feet; usable length may be less.
UNICOM - Aeronautical advisory station ("U" only on WAC)
RP 23,34 - Runways with Right Traffic Patterns (public use) (Not shown on WAC)
RP* - (See Airport/Facility Directory)
VFR Advsy 125.0 - VFR Advisory Service shown where ATIS is not available and frequency is other than primary CT frequency.
WX CAM - Weather Camera (AK)
AOE - Airport of Entry

The symbol L indicates that runway lights are on during hours of darkness. A *L indicates that the pilot must consult the Airport/Facility Directory (A/FD) to determine runway lighting limitations, such as: available on request (by radio call, letter, phone, etc), part-time lighting or pilot/airport controlled lighting. Lighting codes refer to runway edge lights. The lighted runway may not be the longest runway available, and may not be lighted full length. A detailed description of airport and air navigation lighting aids available at each airport can be found in the A/FD. When information is lacking, the respective character is replaced by a dash. The symbol ☆ indicates the existence of a rotating or flashing airport beacon operating continuously sunset to sunrise. The Aeronautical Information Manual (AIM) thoroughly explains the types and uses of airport lighting aids.

CONTROLLED AIRSPACE

Controlled airspace consists of those areas where some or all aircraft may be subject to air traffic control, such as Class A, Class B, Class C, Class D, Class E Surface (SFC) and Class E Airspace.

Class A Airspace within the United States extends from 18,000' up to 60,000' MSL. While visual charts do not depict Class A, it is important to note its existence.

Class B Airspace is shown in abbreviated form on the World Aeronautical Chart (WAC). The Sectional Aeronautical Chart (Sectional) and Terminal Area Chart (TAC) show Class B in greater detail. The MSL ceiling and floor altitudes of each sector are shown in solid blue figures with the last two digits omitted: $\frac{90}{20}$ Radials and arcs used to define Class B are prominently shown on TACs. Detailed rules and requirements associated with the particular Class B are shown. The name by which the Class B is identified is shown as LAS VEGAS CLASS B for example.

Class C Airspace is shown in abbreviated form on WACs. Sectionals and TACs show Class C in greater detail.

The MSL ceiling and floor altitudes of each sector are shown in solid magenta figures with the last two digits eliminated: $\frac{70}{15}$. The following figures identify a sector that extends from the surface to the base of the Class C: $\frac{T}{SFC}$. The name by which the Class C is identified is shown as: BURBANK CLASS C. Separate notes, enclosed in magenta boxes, give the approach control frequencies to be used by arriving VFR aircraft to establish two-way radio communication before entering the Class C (generally within 20 NM):

CTC BURBANK APP WITHIN
20 NM ON 124.6 395.9

Class D Airspace is symbolized by a blue dashed line. Class D operating less than continuous is indicated by the following note: See NOTAMs/Directory for Class D eff hrs. Ceilings of Class D are shown as follows: [30]. A minus in front of the figure is used to indicate "from surface to but not including .... ."

Class E Surface (SFC) Airspace is symbolized by a magenta dashed line. Class E (sfc) operating less than continuous is indicated by the following note: See NOTAMs/Directory for Class E (sfc) eff hrs

Class E Airspace exists at 1200' above ground level unless designated otherwise. The lateral and vertical limits of all Class E up to but not including 18,000' are shown by narrow bands of vignette on Sectionals and TACs. Controlled airspace floors of 700' above the ground are defined by a magenta vignette; floors other than 700' that abut uncontrolled airspace (Class G) are defined by a blue vignette; differing floors greater than 700' above the ground are annotated by a symbol 2400 AGL / 4500 MSL and a number indicating the floor. If the ceiling is less than 18,000' MSL, the value (prefixed by the word "ceiling") is shown along the limits of the controlled airspace. These limits are shown with the same symbol indicated above.

Class E Airspace with floor 700 ft. above surface.

Class E Airspace with floor 1200 ft or greater above surface that abuts Class G Airspace.

UNCONTROLLED AIRSPACE

Class G Airspace within the United States extends up to 14,500' MSL. At and above this altitude is Class E, excluding the airspace less than 1500' above the terrain and certain special use airspace areas.

SPECIAL USE AIRSPACE

Special use airspace confines certain flight activities and restricts entry, or cautions other aircraft operating within specific boundaries. Except for Controlled Firing Areas, special use airspace areas are depicted on visual aeronautical charts. Controlled Firing Areas are not charted because their activities are suspended immediately when spotter aircraft, radar, or ground lookout positions indicate an aircraft might be approaching the area. Nonparticipating aircraft are not required to change their flight paths. Special use airspace areas are shown in their entirety (within the limits of the chart), even when they overlap, adjoin, or when an area is designated within another area. The areas are identified by type and identifying name or number, positioned either within or immediately adjacent to the area.

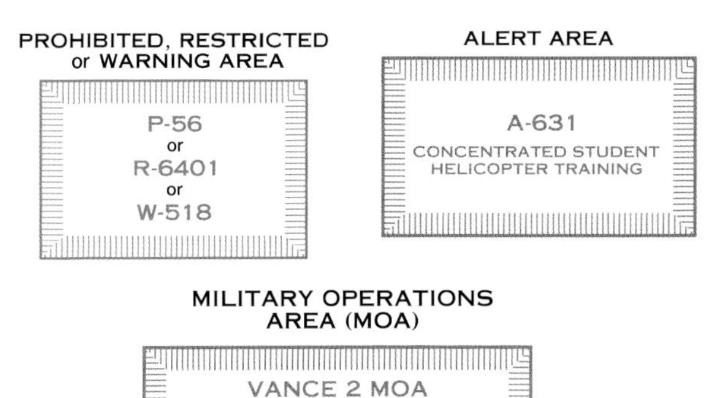

OTHER AIRSPACE AREAS

Mode C Required Airspace (from the surface to 10,000' MSL) within 30 NM radius of the primary airport(s) for which a Class B is designated, is depicted by a solid magenta line MODE C / 30 NM. Mode C is required but not depicted for operations within and above all Class C up to 10,000' MSL. Enroute Mode C requirements (at and above 10,000' MSL except in airspace at and below 2500' AGL) are not depicted. See FAR 91.215 and the AIM.

FAR 93 Airports and heliports where Federal Aviation Regulation (FAR 93) special air traffic rules and airport traffic patterns apply are shown by "boxing" the airport name.

FAR 91 Airports where fixed wing special visual flight rules operations are prohibited (FAR 91) are shown with the type "NO SVFR" above the airport name.

National Security Areas indicated with a broken magenta line and **Special Flight Rules Areas (SFRAs)** indicated with the following symbol: , consist of airspace with defined vertical and lateral dimensions established at locations where there is a requirement for increased security and safety of ground facilities. Pilots are requested to voluntarily avoid flying through these depicted areas. When necessary, flight may be temporarily prohibited.

The Washington DC Flight Restricted Zone (FRZ) is related to National Security. It is depicted using the Prohibited/Restricted/Warning Area symbology and is located within the SFRA. It is defined as the airspace within approximately a 13 to 15 NM radius of the KDCA VOR-DME. Additional requirements are levied upon operators requesting access to operate inside the National Capital Region.

Temporary Flight Restriction (TFR) Areas Relating to National Security are indicated with a broken blue line . A Temporary Flight Restriction (TFR) is a type of Notice to Airmen (NOTAM). A TFR defines an area restricted to air travel due to a hazardous condition, a special event, or a general warning for the entire airspace. The text of the actual TFR contains the fine points of the restriction. It is important to note that only TFRs relating to National Security are charted.

Air Defense Identification Zones (ADIZs) are symbolized using the ADIZ symbol: As defined in 14 CFR Part 99, an ADIZ is an area in which the ready identification, location, and control of all aircraft is required in the interest of national security. ADIZ boundaries include Alaska, Canada and the Contiguous U.S.

Terminal Radar Service Areas (TRSAs) are shown in their entirety, symbolized by a screened black outline of the entire area including the various sectors within the area .

The outer limit of the entire TRSA is a continuous screened black line. The various sectors within the TRSA are symbolized by slightly narrower screened black lines.

Each sector altitude is identified in solid black color by the MSL ceiling and floor values of the respective sector, eliminating the last two digits. A leader line is used when the altitude values must be positioned outside the respective sectors because of space limitations. The TRSA name is shown near the north position of the TRSA as follows: PALM SPRINGS TRSA. Associated frequencies are listed in a table on the chart border.

Military Training Routes (MTRs) are shown on Sectionals and TACs. They are identified by the route designator: ← IR21 . Route designators are shown in solid black on the route centerline, positioned along the route for continuity. The designator IR or VR is not repeated when two or more routes are established over the same airspace, e.g., IR201-205-227. Routes numbered 001 to 099 are shown as IR1 or VR99, eliminating the initial zeros. Direction of flight along the route is indicated by small arrowheads adjacent to and in conjunction with each route designator.

The following note appears on Sectionals and TACs covering the conterminous United States.

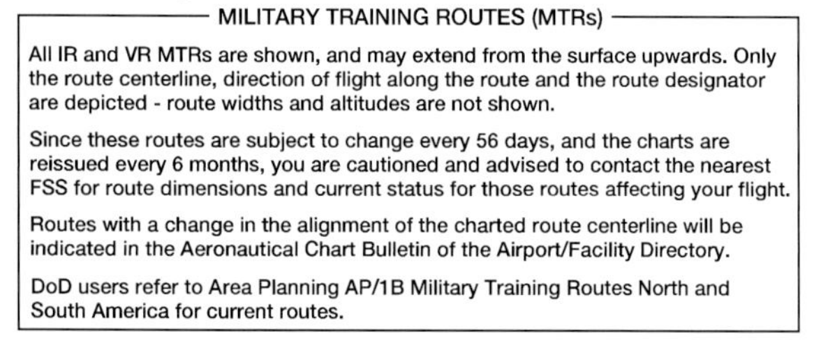
MILITARY TRAINING ROUTES (MTRs)

All IR and VR MTRs are shown, and may extend from the surface upwards. Only the route centerline, direction of flight along the route and the route designator are depicted - route widths and altitudes are not shown.

Since these routes are subject to change every 56 days, and the charts are reissued every 6 months, you are cautioned and advised to contact the nearest FSS for route dimensions and current status for those routes affecting your flight.

Routes with a change in the alignment of the charted route centerline will be indicated in the Aeronautical Chart Bulletin of the Airport/Facility Directory.

DoD users refer to Area Planning AP/1B Military Training Routes North and South America for current routes.

There are IFR (IR) and VFR (VR) routes as follows:

Route identification:

a. Routes at or below 1500' AGL (with no segment above 1500') are identified by four-digit numbers; e.g., VR1007, etc. These routes are generally developed for flight under Visual Flight Rules.

b. Routes above 1500' AGL (some segments of these routes may be below 1500') are identified by three or fewer digit numbers; e.g., IR21, VR302, etc. These routes are developed for flight under Instrument Flight Rules.

MTRs can vary in width from 4 to 16 miles. Detailed route width information is available in the Flight Information Publication (FLIP) AP/1B (a DoD publication), or in the Digital Aeronautical Chart Supplement (DACS) produced by AeroNav Services.

Special Military Activity areas are indicated on the Sectionals by a boxed note in black type. The note contains radio frequency information for obtaining area activity status.

SPECIAL MILITARY ACTIVITY
CTC MOBILE RADIO
ON 123.6
FOR ACTIVITY STATUS

TERMINAL AREA CHART (TAC) COVERAGE

TAC coverage is shown on appropriate Sectionals by a 1/4" masked line as indicated below. Within this area, pilots should use TACs which provide greater detail and clarity of information. A note to this effect appears near the masked boundary line.

LOS ANGELES TERMINAL AREA
Pilots are encouraged to use the Los Angeles VFR Terminal Area Chart for flights at or below 10,000

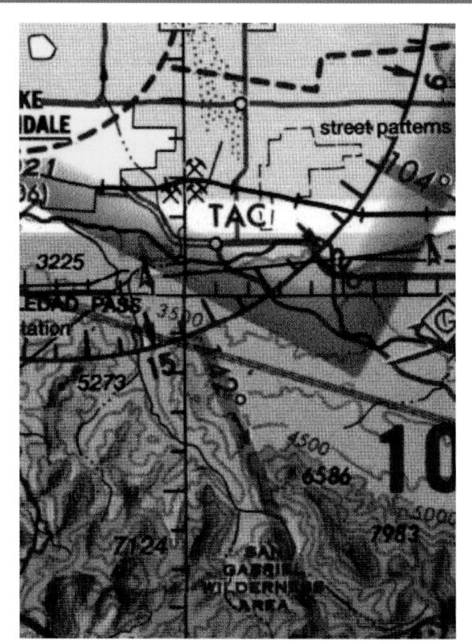

INSET COVERAGE

Inset coverage is shown on appropriate Sectionals by a 1/8" masked line as indicated below. A note to this effect appears near the masked boundary line.

If inset chart is on the same chart as outline:

INDIANAPOLIS INSET
See inset chart for additional detail

If inset chart is on a different chart:

INDIANAPOLIS INSET
See inset chart on the St. Louis
Sectional for additional information

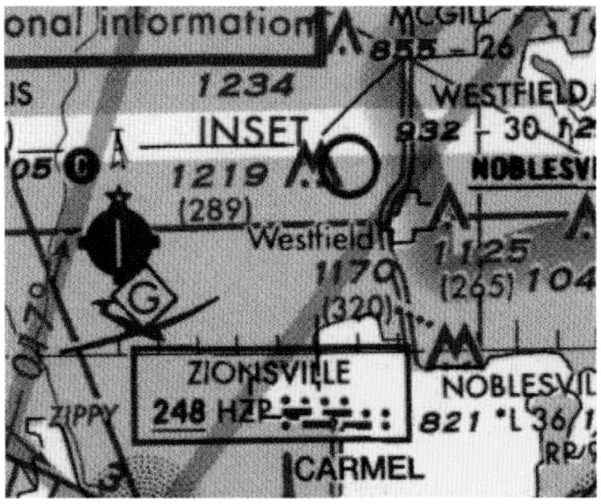

CHART TABULATIONS

Airport Tower Communications are provided in a columnized tabulation for all tower-controlled airports that appear on the respective chart. Airport names are listed alphabetically. If the airport is military, the type of airfield, e.g., AAF, AFB, NAS, is shown after the airfield name. In addition to the airport name, tower operating hours, primary VHF/UHF local Control Tower (CT), Ground Control (GND CON), and Automatic Terminal Information Service (ATIS) frequencies, when available, will be given. An asterisk (*) indicates that the part-time tower frequency is remoted to a colocated full-time FSS for use as Airport Advisory Service (AAS) when the tower is closed. Airport Surveillance Radar (ASR) and/or Precision Approach Radar (PAR) procedures are listed when available.

Approach Control Communications are provided in a columnized tabulation listing Class B, Class C, Terminal Radar Service Areas (TRSA) and Selected Radar Facilities when available. Primary VHF/UHF frequencies are provided for each facility. Sectorization occurs when more than one frequency exists and/or is approach direction dependent. Availability of service hours is also provided.

Special Use Airspace (SUA) Prohibited, Restricted and Warning Areas are presented in blue and listed numerically for U.S. and other countries. Restricted, Danger and Advisory Areas outside the U.S. are tabulated separately in blue. A tabulation of Alert Areas (listed numerically) and Military Operations Areas (MOA) (listed alphabetically) appear on the chart in magenta. All are supplemented with altitude, time of use and the controlling agency/contact facility, and its frequency, when available. The controlling agency will be shown when the contact facility and frequency data is unavailable.

VFR AERONAUTICAL CHARTS

Airports with control towers are indicated on the face of the chart by the letters CT followed by the primary VHF local control frequency (ies). Information for each tower is listed in the table below. Operational hours are local time. The primary VHF and UHF local control frequencies are listed. An asterisk (*) indicates the part-time tower frequency is remoted to a collocated full-time FSS for use as Airport Advisory Service (AAS) during hours the tower is closed. The primary VHF and UHF ground control frequencies are listed.

Automatic Terminal Information Service (ATIS) frequencies shown on the face of the chart are primary arrival VHF/UHF frequencies. All ATIS frequencies are listed in the table below. ATIS operational hours may differ from tower operational hours.

ASR and/or PAR indicate Radar Instrument Approach available.

"MON-FRI" indicates Monday through Friday.

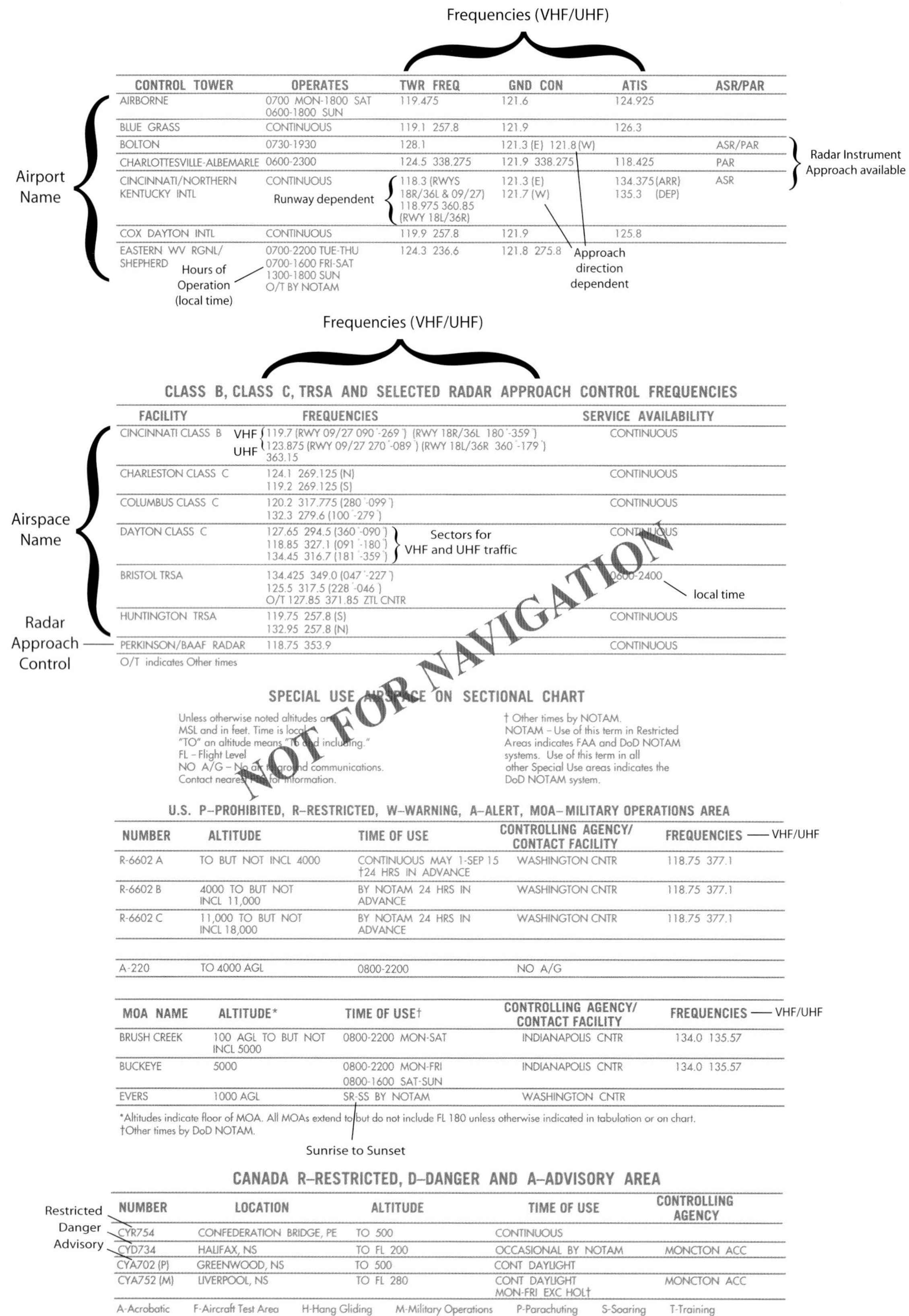

CONTROL TOWER	OPERATES	TWR FREQ	GND CON	ATIS	ASR/PAR
AIRBORNE	0700 MON-1800 SAT 0600-1800 SUN	119.475	121.6	124.925	
BLUE GRASS	CONTINUOUS	119.1 257.8	121.9	126.3	
BOLTON	0730-1930	128.1	121.3 (E) 121.8 (W)		ASR/PAR
CHARLOTTESVILLE-ALBEMARLE	0600-2300	124.5 338.275	121.9 338.275	118.425	PAR
CINCINNATI/NORTHERN KENTUCKY INTL	CONTINUOUS	118.3 (RWYS 18R/36L & 09/27) 118.975 360.85 (RWY 18L/36R)	121.3 (E) 121.7 (W)	134.375 (ARR) 135.3 (DEP)	ASR
COX DAYTON INTL	CONTINUOUS	119.9 257.8	121.9	125.8	
EASTERN WV RGNL/ SHEPHERD	0700-2200 TUE-THU 0700-1600 FRI-SAT 1300-1800 SUN O/T BY NOTAM	124.3 236.6	121.8 275.8		

CLASS B, CLASS C, TRSA AND SELECTED RADAR APPROACH CONTROL FREQUENCIES

FACILITY	FREQUENCIES	SERVICE AVAILABILITY
CINCINNATI CLASS B	119.7 (RWY 09/27 090°-269°) (RWY 18R/36L 180°-359°) 123.875 (RWY 09/27 270°-089°) (RWY 18L/36R 360°-179°) 363.15	CONTINUOUS
CHARLESTON CLASS C	124.1 269.125 (N) 119.2 269.125 (S)	CONTINUOUS
COLUMBUS CLASS C	120.2 317.775 (280°-099°) 132.3 279.6 (100°-279°)	CONTINUOUS
DAYTON CLASS C	127.65 294.5 (360°-090°) 118.85 327.1 (091°-180°) 134.45 316.7 (181°-359°)	CONTINUOUS
BRISTOL TRSA	134.425 349.0 (047°-227°) 125.5 317.5 (228°-046°) O/T 127.85 371.85 ZTL CNTR	0600-2400
HUNTINGTON TRSA	119.75 257.8 (S) 132.95 257.8 (N)	CONTINUOUS
PERKINSON/BAAF RADAR	118.75 353.9	CONTINUOUS

O/T indicates Other times

SPECIAL USE AIRSPACE ON SECTIONAL CHART

Unless otherwise noted altitudes are MSL and in feet. Time is local.
"TO" an altitude means "To and including."
FL – Flight Level
NO A/G – No air to ground communications. Contact nearest FSS for information.

† Other times by NOTAM.
NOTAM – Use of this term in Restricted Areas indicates FAA and DoD NOTAM systems. Use of this term in all other Special Use areas indicates the DoD NOTAM system.

U.S. P–PROHIBITED, R–RESTRICTED, W–WARNING, A–ALERT, MOA–MILITARY OPERATIONS AREA

NUMBER	ALTITUDE	TIME OF USE	CONTROLLING AGENCY/ CONTACT FACILITY	FREQUENCIES
R-6602 A	TO BUT NOT INCL 4000	CONTINUOUS MAY 1-SEP 15 †24 HRS IN ADVANCE	WASHINGTON CNTR	118.75 377.1
R-6602 B	4000 TO BUT NOT INCL 11,000	BY NOTAM 24 HRS IN ADVANCE	WASHINGTON CNTR	118.75 377.1
R-6602 C	11,000 TO BUT NOT INCL 18,000	BY NOTAM 24 HRS IN ADVANCE	WASHINGTON CNTR	118.75 377.1
A-220	TO 4000 AGL	0800-2200	NO A/G	

MOA NAME	ALTITUDE*	TIME OF USE†	CONTROLLING AGENCY/ CONTACT FACILITY	FREQUENCIES
BRUSH CREEK	100 AGL TO BUT NOT INCL 5000	0800-2200 MON-SAT	INDIANAPOLIS CNTR	134.0 135.57
BUCKEYE	5000	0800-2200 MON-FRI 0800-1600 SAT-SUN	INDIANAPOLIS CNTR	134.0 135.57
EVERS	1000 AGL	SR-SS BY NOTAM	WASHINGTON CNTR	

*Altitudes indicate floor of MOA. All MOAs extend to but do not include FL 180 unless otherwise indicated in tabulation or on chart.
†Other times by DoD NOTAM.

CANADA R–RESTRICTED, D–DANGER AND A–ADVISORY AREA

NUMBER	LOCATION	ALTITUDE	TIME OF USE	CONTROLLING AGENCY
CYR754	CONFEDERATION BRIDGE, PE	TO 500	CONTINUOUS	
CYD734	HALIFAX, NS	TO FL 200	OCCASIONAL BY NOTAM	MONCTON ACC
CYA702 (P)	GREENWOOD, NS	TO 500	CONT DAYLIGHT	
CYA752 (M)	LIVERPOOL, NS	TO FL 280	CONT DAYLIGHT MON-FRI EXC HOL†	MONCTON ACC

A-Acrobatic F-Aircraft Test Area H-Hang Gliding M-Military Operations P-Parachuting S-Soaring T-Training

VFR AERONAUTICAL CHART SYMBOLS

AERONAUTICAL INFORMATION

TOPOGRAPHIC INFORMATION

CULTURE

HYDROGRAPHY

RELIEF

HELICOPTER ROUTE CHARTS

VFR FLYWAY PLANNING CHARTS

GENERAL INFORMATION

Symbols shown are for World Aeronautical Charts (WACs), Sectional Aeronautical Charts (Sectionals), Terminal Area Charts (TACs), VFR Flyway Planning Charts and Helicopter Route Charts. When a symbol is different on any VFR chart series, it will be annotated, e.g., “WAC” or “Not shown on WAC”.

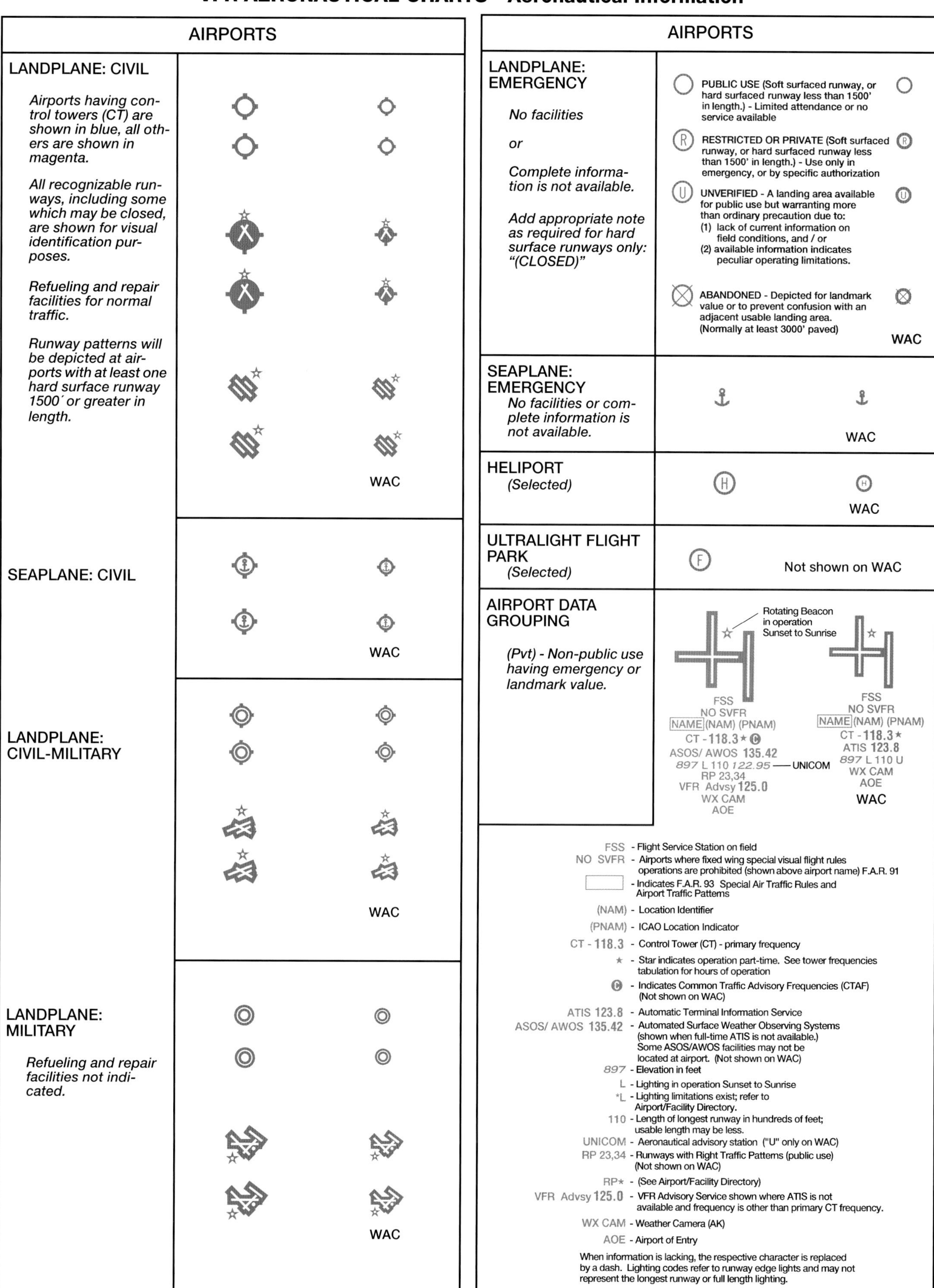
AIRPORTS
LANDPLANE: CIVIL
Airports having control towers (CT) are shown in blue, all others are shown in magenta.
All recognizable runways, including some which may be closed, are shown for visual identification purposes.
Refueling and repair facilities for normal traffic.
Runway patterns will be depicted at airports with at least one hard surface runway 1500′ or greater in length.
WAC
SEAPLANE: CIVIL
WAC
LANDPLANE: CIVIL-MILITARY
WAC
LANDPLANE: MILITARY
Refueling and repair facilities not indicated.
WAC
AIRPORTS
LANDPLANE: EMERGENCY
No facilities
or
Complete information is not available.
Add appropriate note as required for hard surface runways only: "(CLOSED)"
PUBLIC USE (Soft surfaced runway, or hard surfaced runway less than 1500' in length.) - Limited attendance or no service available
RESTRICTED OR PRIVATE (Soft surfaced runway, or hard surfaced runway less than 1500' in length.) - Use only in emergency, or by specific authorization
UNVERIFIED - A landing area available for public use but warranting more than ordinary precaution due to:
(1) lack of current information on field conditions, and / or
(2) available information indicates peculiar operating limitations.
ABANDONED - Depicted for landmark value or to prevent confusion with an adjacent usable landing area. (Normally at least 3000' paved)
WAC
SEAPLANE: EMERGENCY
No facilities or complete information is not available.
WAC
HELIPORT
(Selected)
WAC
ULTRALIGHT FLIGHT PARK
(Selected)
Not shown on WAC
AIRPORT DATA GROUPING
(Pvt) - Non-public use having emergency or landmark value.
Rotating Beacon in operation Sunset to Sunrise
FSS
NO SVFR
NAME (NAM) (PNAM)
CT - 118.3 ★ C
ASOS/ AWOS 135.42
897 L 110 122.95 — UNICOM
RP 23,34
VFR Advsy 125.0
WX CAM
AOE
FSS
NO SVFR
NAME (NAM) (PNAM)
CT - 118.3 ★
ATIS 123.8
897 L 110 U
WX CAM
AOE
WAC
FSS - Flight Service Station on field
NO SVFR - Airports where fixed wing special visual flight rules operations are prohibited (shown above airport name) F.A.R. 91
- Indicates F.A.R. 93 Special Air Traffic Rules and Airport Traffic Patterns
(NAM) - Location Identifier
(PNAM) - ICAO Location Indicator
CT - 118.3 - Control Tower (CT) - primary frequency
★ - Star indicates operation part-time. See tower frequencies tabulation for hours of operation
C - Indicates Common Traffic Advisory Frequencies (CTAF) (Not shown on WAC)
ATIS 123.8 - Automatic Terminal Information Service
ASOS/ AWOS 135.42 - Automated Surface Weather Observing Systems (shown when full-time ATIS is not available.) Some ASOS/AWOS facilities may not be located at airport. (Not shown on WAC)
897 - Elevation in feet
L - Lighting in operation Sunset to Sunrise
*L - Lighting limitations exist; refer to Airport/Facility Directory.
110 - Length of longest runway in hundreds of feet; usable length may be less.
UNICOM - Aeronautical advisory station ("U" only on WAC)
RP 23,34 - Runways with Right Traffic Patterns (public use) (Not shown on WAC)
RP* - (See Airport/Facility Directory)
VFR Advsy 125.0 - VFR Advisory Service shown where ATIS is not available and frequency is other than primary CT frequency.
WX CAM - Weather Camera (AK)
AOE - Airport of Entry
When information is lacking, the respective character is replaced by a dash. Lighting codes refer to runway edge lights and may not represent the longest runway or full length lighting.

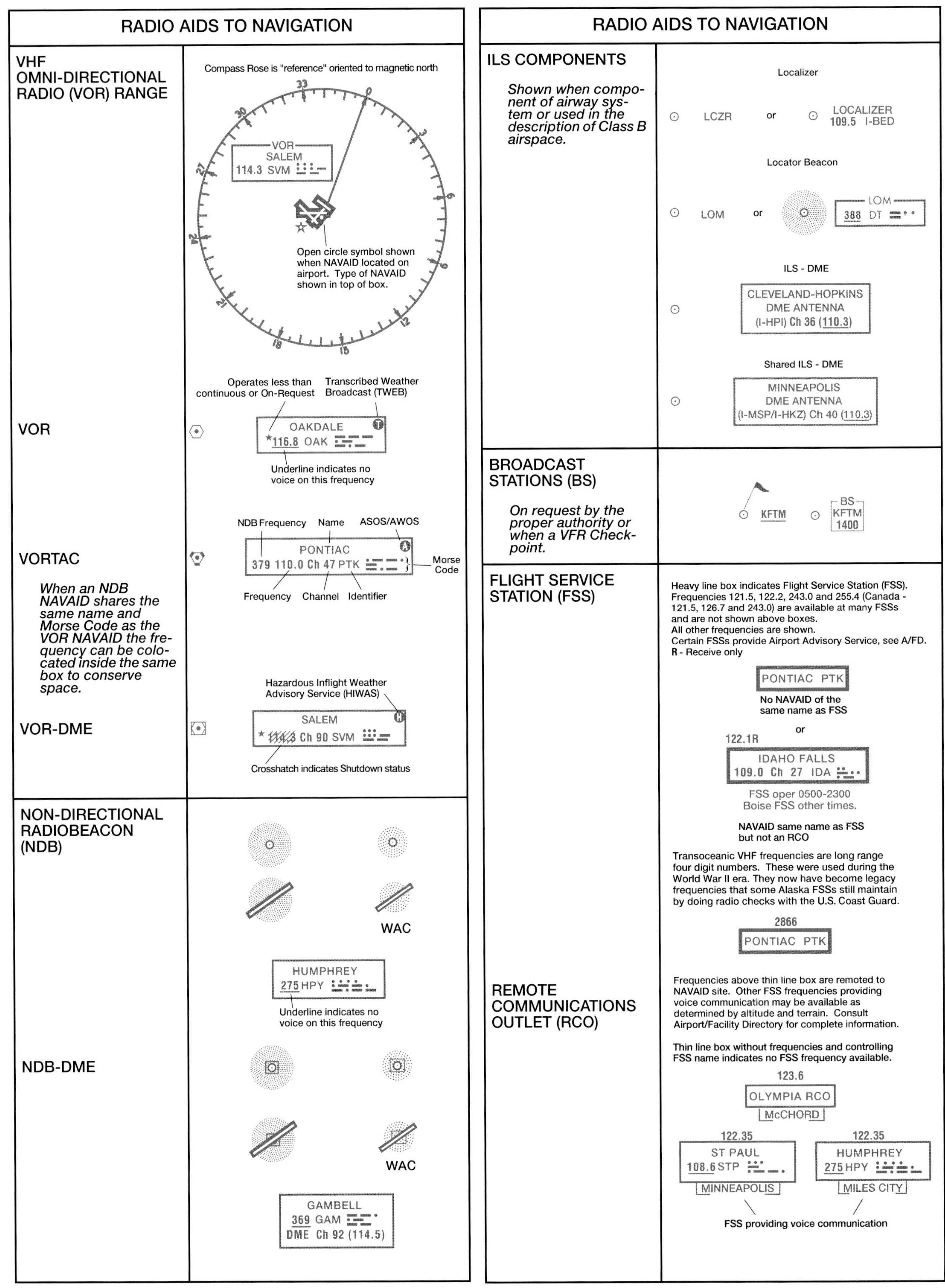
RADIO AIDS TO NAVIGATION
VHF OMNI-DIRECTIONAL RADIO (VOR) RANGE
Compass Rose is "reference" oriented to magnetic north
VOR SALEM 114.3 SVM
Open circle symbol shown when NAVAID located on airport. Type of NAVAID shown in top of box.
Operates less than continuous or On-Request
Transcribed Weather Broadcast (TWEB)
VOR
OAKDALE
*116.8 OAK
Underline indicates no voice on this frequency
NDB Frequency
Name
ASOS/AWOS
VORTAC
PONTIAC
379 110.0 Ch 47 PTK
Morse Code
Frequency
Channel
Identifier
When an NDB NAVAID shares the same name and Morse Code as the VOR NAVAID the frequency can be colocated inside the same box to conserve space.
Hazardous Inflight Weather Advisory Service (HIWAS)
VOR-DME
SALEM
* 114.3 Ch 90 SVM
Crosshatch indicates Shutdown status
NON-DIRECTIONAL RADIOBEACON (NDB)
WAC
HUMPHREY
275 HPY
Underline indicates no voice on this frequency
NDB-DME
WAC
GAMBELL
369 GAM
DME Ch 92 (114.5)
RADIO AIDS TO NAVIGATION
ILS COMPONENTS
Shown when component of airway system or used in the description of Class B airspace.
Localizer
LCZR
or
LOCALIZER
109.5 I-BED
Locator Beacon
LOM
or
LOM
388 DT
ILS - DME
CLEVELAND-HOPKINS
DME ANTENNA
(I-HPI) Ch 36 (110.3)
Shared ILS - DME
MINNEAPOLIS
DME ANTENNA
(I-MSP/I-HKZ) Ch 40 (110.3)
BROADCAST STATIONS (BS)
On request by the proper authority or when a VFR Checkpoint.
KFTM
BS
KFTM
1400
FLIGHT SERVICE STATION (FSS)
Heavy line box indicates Flight Service Station (FSS). Frequencies 121.5, 122.2, 243.0 and 255.4 (Canada - 121.5, 126.7 and 243.0) are available at many FSSs and are not shown above boxes.
All other frequencies are shown.
Certain FSSs provide Airport Advisory Service, see A/FD.
R - Receive only
PONTIAC PTK
No NAVAID of the same name as FSS
or
122.1R
IDAHO FALLS
109.0 Ch 27 IDA
FSS oper 0500-2300
Boise FSS other times.
NAVAID same name as FSS but not an RCO
Transoceanic VHF frequencies are long range four digit numbers. These were used during the World War II era. They now have become legacy frequencies that some Alaska FSSs still maintain by doing radio checks with the U.S. Coast Guard.
2866
PONTIAC PTK
REMOTE COMMUNICATIONS OUTLET (RCO)
Frequencies above thin line box are remoted to NAVAID site. Other FSS frequencies providing voice communication may be available as determined by altitude and terrain. Consult Airport/Facility Directory for complete information.
Thin line box without frequencies and controlling FSS name indicates no FSS frequency available.
123.6
OLYMPIA RCO
McCHORD
122.35
ST PAUL
108.6 STP
MINNEAPOLIS
122.35
HUMPHREY
275 HPY
MILES CITY
FSS providing voice communication

RADIO AIDS TO NAVIGATION

AIR FORCE STATION (AFS)	122.0 AFS 123.6 POINT BARROW; 122.4 AFS 123.6 CAPE LEWISTON 206 LWS — AFS at airport with NDB
LONG RANGE RADAR STATION (LRRS)	122.4 LRRS 122.55 BARTER ISLAND; 122.4 LRRS 123.6 CAPE LISBURNE 385 LUR — LRRS at airport with NDB
OFF AIRPORT AWOS/ASOS	SANDBERG ASOS 120.625
ALASKA WEATHER CAMERA Stand-Alone	ANCHORAGE WX CAM
Colocated with Airport *Must be within 2NM to have same name.*	WRANGELL (68A) 00 - 90 122.6 WX CAM AOE

AIRSPACE INFORMATION

CLASS B AIRSPACE *Appropriate notes as required may be shown.* *Only the airspace effective below 18,000 feet MSL are shown.* *(Mode C see FAR 91.215 /AIM)* *All mileages are nautical (NM).* *All radials are magnetic.*	LAS VEGAS CLASS B LAS 20 NM — NAVAID identifier and distance from facility (TAC Only). LAS 002° — NAVAID identifier and radial from facility (TAC Only). Outer limit only, segments not shown — WAC FOR FLIGHTS AT AND BELOW 8000 MSL SEE KANSAS CITY VFR TERMINAL AREA CHART — WAC only
	80 - Ceiling of Class B in hundreds of feet MSL 40 - Floor of Class B in hundreds of feet MSL CTC LAS VEGAS APP ON 121.1 OR 257.8 — TAC only

AIRSPACE INFORMATION

CLASS C AIRSPACE *Appropriate notes as required may be shown.* *(Mode C see FAR 91.215 /AIM)*	BURBANK CLASS C See NOTAMs/Directory for Class C eff hrs BOISE CLASS C See NOTAMs/Directory for Class C eff hrs Outer limit only, segments not shown — WAC FOR FLIGHTS AT OR BELOW 6600 MSL SEE PHOENIX VFR SECTIONAL CHART — WAC only
	48 - Ceiling of Class C in hundreds of feet MSL 30 - Floor of Class C in hundreds of feet MSL T - Ceiling is to but not including floor of Class B SFC - Surface CTC BURBANK APP WITHIN 20 NM ON 124.6 395.9 — Not shown on WAC
CLASS D AIRSPACE	See NOTAMs/Directory for Class D eff hrs — 27 See NOTAMs/Directory for Class D/E (sfc) eff hrs — -27 (A minus in front of the figure is used to indicate "from surface to but not including...") ALTITUDE IN HUNDREDS OF FEET MSL Not shown on WAC

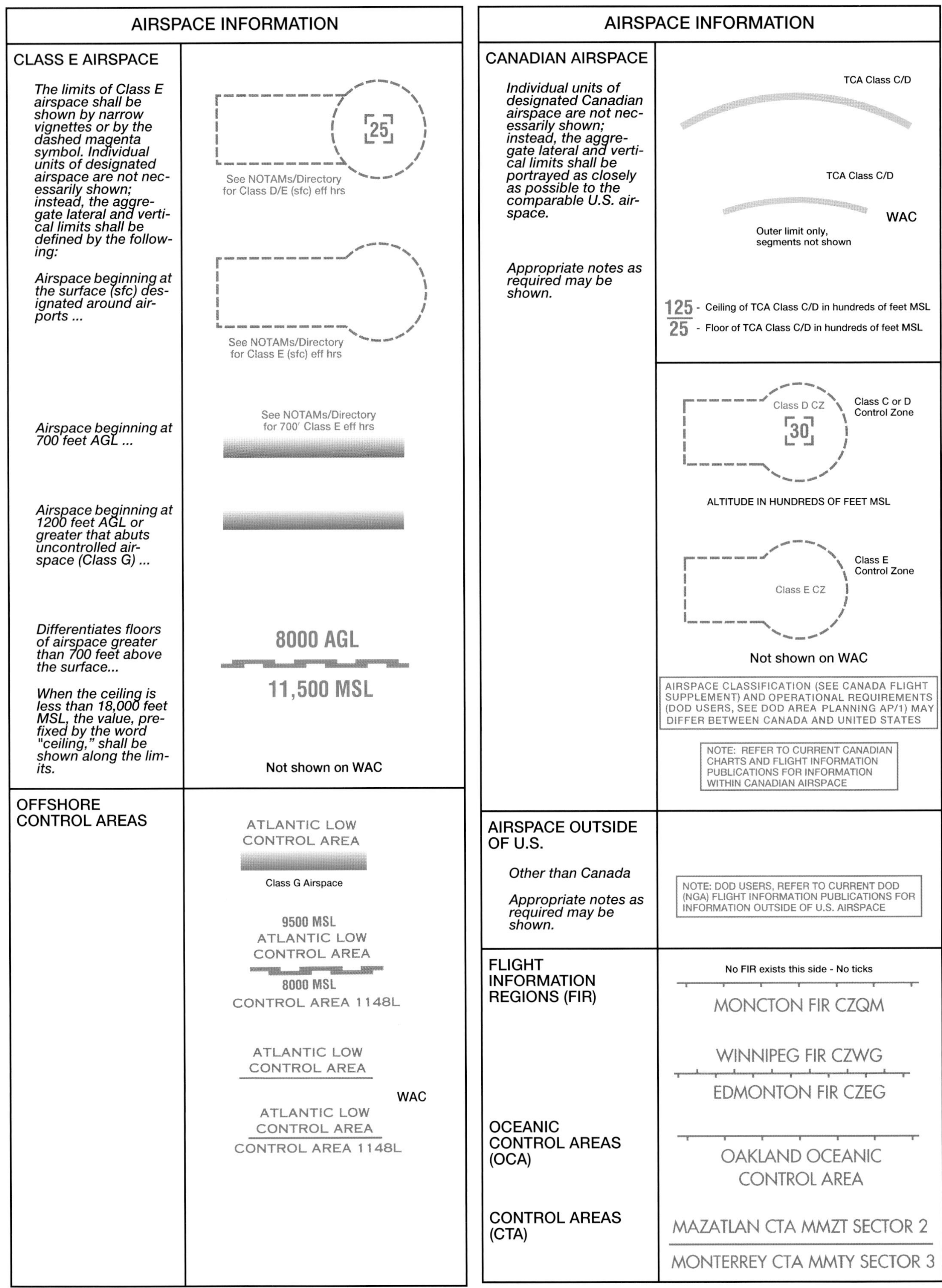
AIRSPACE INFORMATION
CLASS E AIRSPACE
The limits of Class E airspace shall be shown by narrow vignettes or by the dashed magenta symbol. Individual units of designated airspace are not necessarily shown; instead, the aggregate lateral and vertical limits shall be defined by the following:
Airspace beginning at the surface (sfc) designated around airports ...
25
See NOTAMs/Directory for Class D/E (sfc) eff hrs
See NOTAMs/Directory for Class E (sfc) eff hrs
Airspace beginning at 700 feet AGL ...
See NOTAMs/Directory for 700′ Class E eff hrs
Airspace beginning at 1200 feet AGL or greater that abuts uncontrolled airspace (Class G) ...
Differentiates floors of airspace greater than 700 feet above the surface...
When the ceiling is less than 18,000 feet MSL, the value, prefixed by the word "ceiling," shall be shown along the limits.
8000 AGL
11,500 MSL
Not shown on WAC
OFFSHORE CONTROL AREAS
ATLANTIC LOW CONTROL AREA
Class G Airspace
9500 MSL
ATLANTIC LOW CONTROL AREA
8000 MSL
CONTROL AREA 1148L
ATLANTIC LOW CONTROL AREA
WAC
ATLANTIC LOW CONTROL AREA
CONTROL AREA 1148L
AIRSPACE INFORMATION
CANADIAN AIRSPACE
Individual units of designated Canadian airspace are not necessarily shown; instead, the aggregate lateral and vertical limits shall be portrayed as closely as possible to the comparable U.S. airspace.
Appropriate notes as required may be shown.
TCA Class C/D
TCA Class C/D
WAC
Outer limit only, segments not shown
125 - Ceiling of TCA Class C/D in hundreds of feet MSL
25 - Floor of TCA Class C/D in hundreds of feet MSL
Class D CZ
30
Class C or D Control Zone
ALTITUDE IN HUNDREDS OF FEET MSL
Class E CZ
Class E Control Zone
Not shown on WAC
AIRSPACE CLASSIFICATION (SEE CANADA FLIGHT SUPPLEMENT) AND OPERATIONAL REQUIREMENTS (DOD USERS, SEE DOD AREA PLANNING AP/1) MAY DIFFER BETWEEN CANADA AND UNITED STATES
NOTE: REFER TO CURRENT CANADIAN CHARTS AND FLIGHT INFORMATION PUBLICATIONS FOR INFORMATION WITHIN CANADIAN AIRSPACE
AIRSPACE OUTSIDE OF U.S.
Other than Canada
Appropriate notes as required may be shown.
NOTE: DOD USERS, REFER TO CURRENT DOD (NGA) FLIGHT INFORMATION PUBLICATIONS FOR INFORMATION OUTSIDE OF U.S. AIRSPACE
FLIGHT INFORMATION REGIONS (FIR)
No FIR exists this side - No ticks
MONCTON FIR CZQM
WINNIPEG FIR CZWG
EDMONTON FIR CZEG
OCEANIC CONTROL AREAS (OCA)
OAKLAND OCEANIC CONTROL AREA
CONTROL AREAS (CTA)
MAZATLAN CTA MMZT SECTOR 2
MONTERREY CTA MMTY SECTOR 3

AIRSPACE INFORMATION

LOW ALTITUDE AIRWAYS VOR and LF / MF (CLASS E AIRSPACE)

Low altitude Federal Airways are indicated by centerline.

Only the controlled airspace effective below 18,000 feet MSL is shown.

V2N ← 270°
Alternate Airway radial

Total mileage between NAVAIDs on direct Airways.

25
V2 ← 255°
Enroute Airway radial

R40
LF / MF Airway

V2N ← 270°
Alternate Airway radial

V2 ← 255°
Enroute Airway radial

R40
LF / MF Airway

WAC

MISCELLANEOUS AIR ROUTES

Combined Federal Airway/RNAV "T" Routes are identified in solid blue type adjacent to the solid magenta federal airway identification. The joint route symbol is screened magenta.

BR 63V ← 265°
Bahama Route

T 319
RNAV Route

A 301
Oceanic & ATS Route

AR5
Atlantic Route

GULF RTE 26
Gulf Route

B ROUTE 2
Class G Route

A 301 T 319
Federal / RNAV Route

BR 63V ← 265°
Bahama Route

T 319
RNAV Route

A 301
Oceanic & ATS Route

AR5
Atlantic Route

GULF RTE 26
Gulf Route

B ROUTE 2
Class G Route

A 301 T 319
Federal / RNAV Route

WAC

AIRSPACE INFORMATION

SPECIAL USE AIRSPACE

Only the airspace effective below 18,000 feet MSL is shown.

The type of area shall be spelled out in large areas if space permits.

P-56
or
R-6401
or
W-518

PROHIBITED, RESTRICTED or WARNING AREA

ALERT AREA
A-631
CONCENTRATED STUDENT HELICOPTER TRAINING

ALERT AREA

VANCE 2 MOA

MILITARY OPERATIONS AREA (MOA)

MILITARY TRAINING ROUTES (MTR)

←IR292

Not shown on WAC

SPECIAL MILITARY ACTIVITY ROUTES (SMAR)

40 / 05 AGL

45 / 05 AGL

Boxed notes shown adjacent to route.

SPECIAL MILITARY ACTIVITY
CTC MOBILE RADIO
ON 123.6
FOR ACTIVITY STATUS

40 — Ceiling of SMAR in hundreds of feet MSL
05 AGL — Floor of SMAR in hundreds of feet AGL

Not shown on WAC

AIRSPACE INFORMATION	
SPECIAL AIR TRAFFIC RULES / AIRPORT PATTERNS (FAR 93) *Appropriate boxed note as required shown adjacent to area.*	SPECIAL NOTICE Pilots are required to obtain an ATC clearance prior to entering this area.
SPACE OPERATIONS AREA (FAR 91.143)	DARKER TINT IS FAR 91.143 AREA Not shown on WAC
MODE C (FAR 91.215) *Appropriate notes as required may be shown.*	MODE C 30 NM
MISCELLANEOUS AIRSPACE AREAS Parachute Jumping Area with Frequency Glider Operating Area Ultralight Activity Hang Glider Activity	122.9 G U H Not shown on WAC
SPECIAL CONSERVATION AREAS National Park, Wildlife Refuge, Primitive and Wilderness Areas, etc.	PAHRANAGAT NATIONAL WILDLIFE REFUGE Not shown on WAC

AIRSPACE INFORMATION	
SPECIAL AIRSPACE AREAS SPECIAL FLIGHT RULES AREA (SFRA) Example: Washington DC *Appropriate notes as required may be shown.* *Note. Delimiting line not shown when it coincides with International Boundary, projection lines or other linear features.*	**Washington DC Metropolitan Area Special Flight Rules Area/Flight Restricted Zone restrictions are in effect.** Special regulations apply to all aircraft operations from the surface to but not including Flight Level 180 in the Washington DC Metropolitan Area. Pilots should contact a local FSS for NOTAM information prior to flight in the Washington DC Metropolitan Area.
FLIGHT RESTRICTED ZONE (FRZ) RELATING TO NATIONAL SECURITY Example: Washington DC	
TEMPORARY FLIGHT RESTRICTION (TFR) RELATING TO NATIONAL SECURITY Example: P-40/R-4009 *Appropriate notes as required may be shown.*	CAUTION P-40 AND R-4009 EXPANDED BY TEMPORARY FLIGHT RESTRICTION. CONTACT AFSS FOR LATEST STATUS AND NOTAMS Not shown on WAC
AIR DEFENSE IDENTIFICATION ZONE (ADIZ) *Note. Delimiting line not shown when it coincides with International Boundary, projection lines or other linear features.*	CONTIGUOUS U.S. ADIZ

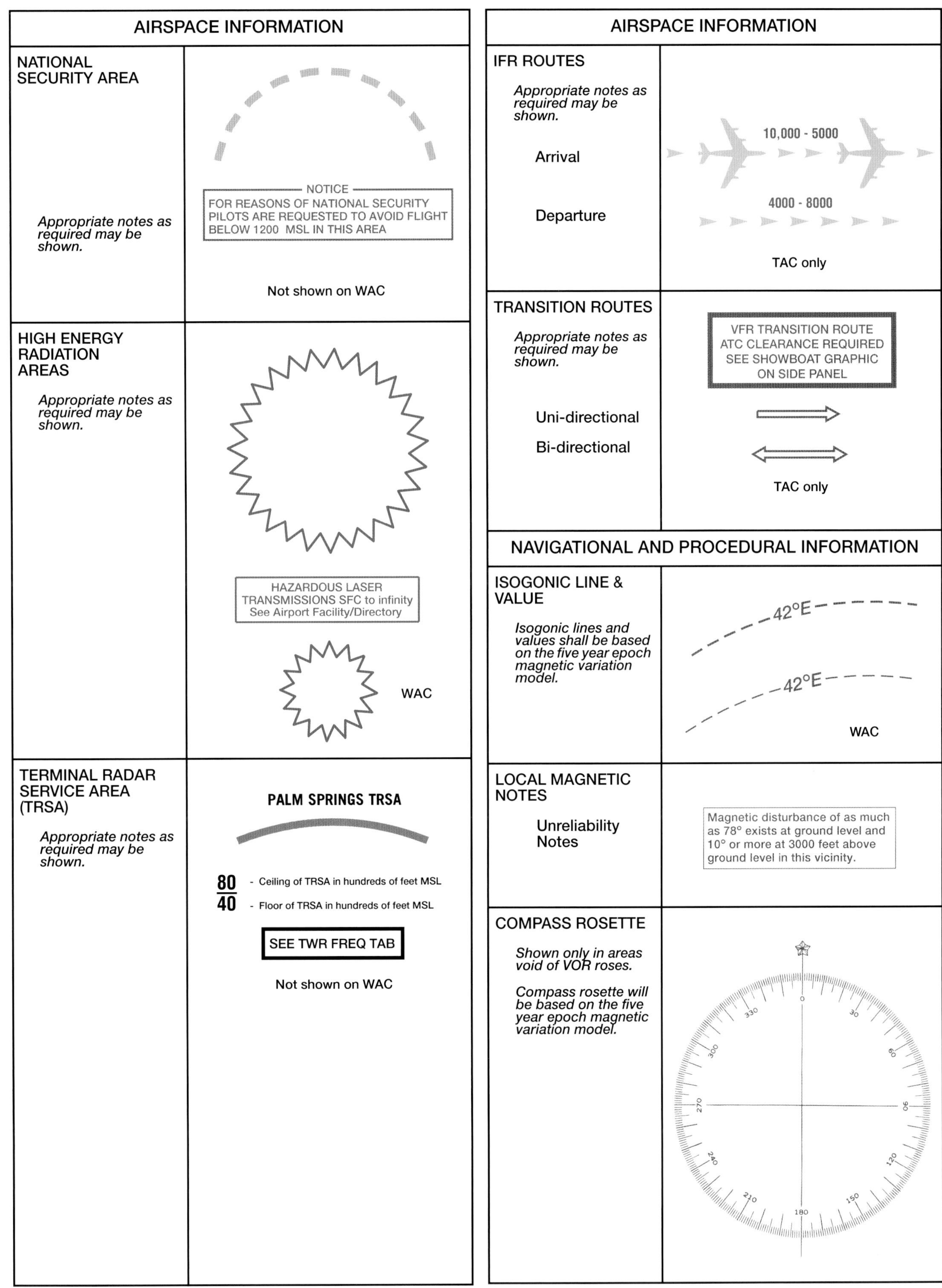
AIRSPACE INFORMATION
NATIONAL SECURITY AREA
Appropriate notes as required may be shown.
NOTICE
FOR REASONS OF NATIONAL SECURITY PILOTS ARE REQUESTED TO AVOID FLIGHT BELOW 1200 MSL IN THIS AREA
Not shown on WAC
HIGH ENERGY RADIATION AREAS
Appropriate notes as required may be shown.
HAZARDOUS LASER TRANSMISSIONS SFC to infinity See Airport Facility/Directory
WAC
TERMINAL RADAR SERVICE AREA (TRSA)
Appropriate notes as required may be shown.
PALM SPRINGS TRSA
80 - Ceiling of TRSA in hundreds of feet MSL
40 - Floor of TRSA in hundreds of feet MSL
SEE TWR FREQ TAB
Not shown on WAC
AIRSPACE INFORMATION
IFR ROUTES
Appropriate notes as required may be shown.
Arrival
10,000 - 5000
Departure
4000 - 8000
TAC only
TRANSITION ROUTES
Appropriate notes as required may be shown.
VFR TRANSITION ROUTE ATC CLEARANCE REQUIRED SEE SHOWBOAT GRAPHIC ON SIDE PANEL
Uni-directional
Bi-directional
TAC only
NAVIGATIONAL AND PROCEDURAL INFORMATION
ISOGONIC LINE & VALUE
Isogonic lines and values shall be based on the five year epoch magnetic variation model.
42°E
42°E
WAC
LOCAL MAGNETIC NOTES
Unreliability Notes
Magnetic disturbance of as much as 78° exists at ground level and 10° or more at 3000 feet above ground level in this vicinity.
COMPASS ROSETTE
Shown only in areas void of VOR roses.
Compass rosette will be based on the five year epoch magnetic variation model.
0
30
60
90
120
150
180
210
240
270
300
330

VFR AERONAUTICAL CHARTS - Aeronautical Information

NAVIGATIONAL AND PROCEDURAL INFORMATION

INTERSECTIONS

Named intersections used as reporting points. Arrows are directed toward facilities which establish intersection.

ANGOO VHF

ROAMS LF / MF

WATSY Combined VHF - LF / MF

Not shown on WAC

AERONAUTICAL LIGHTS

Rotating or Oscillating

Located at Aerodrome

2520 2520

In isolated location on top of high structure

WAC

AERONAUTICAL LIGHTS

Rotating Light with Flashing Code Identification Light

Rotating Light with Course Lights and Site Number

5 Site # 5

18 18

4B 4B

Flashing Light

Fl Fl

Fl Fl

WAC

NAVIGATIONAL AND PROCEDURAL INFORMATION

MARINE LIGHTS

With Characteristics of Light

Oc R SEC Land Light

Al Land Light

WAC

R	Red
*W	White
G	Green
B	Blue
SEC	Sector
F	Fixed
Oc	Single Occulting
Oc (2)	Group Occulting
Oc (2+1)	Composite Group Occulting
Iso	Isophase
Fl	Flashing
Fl (2)	Group Flashing
Fl (2+1)	Composite Group Flashing
Q	Quick
IQ	Interrupted Quick
Mo (A)	Morse Code
FFl	Fixed and Flashing
Al	Alternating
Gp	Group
LFl	Long Flash
Q (3)	Group Quick Flashing
IQ	Interrupted Quick Flashing
VQ	Very Quick Flashing
VQ (3)	Group Very Quick Flashing
IVQ	Interrupted Very Quick Flashing
UQ	Ultra Quick Flashing
IUQ	Interrupted Ultra Quick Flashing

*Marine Lights are white unless otherwise noted. Alternating lights are red and white unless otherwise noted.

VISUAL GROUND SIGNS

Shore and Landmarkers

A33

Arrow points to location of marker

M

Actual location of ground sign

VFR CHECKPOINTS

Pictorial STATE CAPITOL

SIGNAL HILL

NORTHBROOK 113.0 Ch 77 OBK

LEWIS (Pvt) 989 - 27

Not shown on WAC

VFR WAYPOINTS

RNAV — GRANT

Stand-Alone — VPXYZ

Colocated with Visual Checkpoint — NAME (VPXYZ)

Not shown on WAC

NAVIGATIONAL AND PROCEDURAL INFORMATION	
OBSTRUCTION	1473 (394) bldg — Less than 1000′ AGL — 1158 (553) stack 628 UC — Under Construction or reported and position / elevation unverified — 507 UC 3368 (1529) — 1000′ AGL and higher — 2967 (1697) WAC
GROUP OBSTRUCTION	1062 (227) — Less than 1000′ AGL — 1524 (567) 4977 (1432) — 1000′ AGL and higher — 3483 (1634) 2889 (1217) — At least two in group over 1000′ AGL — 4892 (1573) WAC
HIGH-INTENSITY OBSTRUCTION LIGHTS *High-intensity lights may operate part-time or by proximity activation.*	Less than 1000′ AGL 1000′ AGL and higher Group Obstruction WAC
WINDMILL FARMS *When highest wind-mill is unverified, UC will be shown after MSL value.*	CAUTION NUMEROUS WINDMILLS HIGHEST 3624′ MSL UC CAUTION NUMEROUS WINDMILLS HIGHEST 3624′ MSL WAC
MAXIMUM ELEVATION FIGURE (MEF) *(see page 2 for expla-nation).*	13⁵
WARNING AND CAUTION NOTES *Used when specific area is not demar-cated.*	WARNING Extensive fleet and air operations being conducted in offshore areas to approximately 100 miles seaward. CAUTION: Be prepared for loss of horizontal reference at low altitude over lake during hazy conditions and at night.

CHART LIMITS	
OUTLINE ON SECTIONAL OF TERMINAL AREA CHART	LOS ANGELES TERMINAL AREA Pilots are encouraged to use the Los Angeles VFR Terminal Area Chart for flights at or below 10,000′ Not shown on WAC
OUTLINE ON SECTIONAL OF INSET CHART	If inset chart is on a different chart: INDIANAPOLIS INSET See inset chart on the St. Louis Sectional for additional information If inset chart is on the same chart as outline: INDIANAPOLIS INSET See inset chart for additional detail Not shown on WAC

VFR AERONAUTICAL CHARTS - Topographic Information

CULTURE	
RAILROADS *All gauges* Single Track	WAC
Double Track	WAC
More Than Two Tracks	3 tracks
Electric	electric
RAILROADS IN JUXTAPOSITION	
RAILROAD-NONOPERATING, ABANDONED, DESTROYED OR UNDER CONSTRUCTION	abandoned
RAILROAD YARDS Limiting Track To Scale	railroad yard
Location Only	railroad yard
RAILROAD STATIONS	station station

CULTURE	
RAILROAD SIDINGS AND SHORT SPURS	
ROADS Dual-Lane Divided Highway Category 1	WAC
Primary Category 2	WAC
Secondary Category 2	
TRAILS Category 3 *Provides symbolization for dismantled railroad when combined with label "dismantled railroad."*	
ROAD MARKERS Interstate Route No.	80
U.S. Route No.	40
Air Marked Identification Label	13
ROAD NAMES	LINCOLN HIGHWAY LINCOLN HIGHWAY WAC
ROADS UNDER CONSTRUCTION	under construction

CULTURE	
BRIDGES AND VIADUCTS	
Railroad	
Road	
OVERPASSES AND UNDERPASSES	
CAUSEWAYS	
TUNNELS-ROAD AND RAILROAD	
POPULATED PLACES	
Large Cities Category 1	
Cities and Large Towns Category 2	
POPULATED PLACES	
Towns and Villages Category 3	WAC

CULTURE	
FERRIES, FERRY SLIPS AND FORDS	ford ferry ford ferry ferry ferry slip
PROMINENT FENCES	
BOUNDARIES	
International	
State or Province	
Convention or Mandate Line	RUSSIA UNITED STATES
Date Line	INTERNATIONAL (Monday) DATE LINE (Sunday)

CULTURE	
TIME ZONES	PST +8 (+7DT) = UTC MST +7 (+6DT) = UTC Not shown on WAC
MINES OR QUARRIES *Shaft Mines or Quarries*	
POWER TRANSMISSION & TELECOMMUNICA-TION LINES	WAC
PIPELINES	pipeline
Underground	underground pipeline
DAMS	
DAM CARRYING ROAD	
PASSABLE LOCKS	locks

CULTURE	
SMALL LOCKS	
WEIRS AND JETTIES	jetties
SEAWALLS	seawall
BREAKWATERS	breakwater breakwater
PIERS, WHARFS, QUAYS, ETC.	piers piers
MISCELLANEOUS CULTURAL FEATURES	stadium fort cemetery
OUTDOOR THEATER	
WELLS Other Than Water	oil

CULTURE

RACE TRACKS	
LOOKOUT TOWERS Air marked identification	618 (Elevation Base of Tower)
LANDMARK AREAS	landfill
TANKS	water oil gas
COAST GUARD STATION	CG
AERIAL CABLEWAYS, CONVEYORS, ETC.	aerial cableway — aerial cableway WAC

HYDROGRAPHY

OPEN WATER	
INLAND WATER	
OPEN / INLAND WATER	

HYDROGRAPHY

SHORELINES Definite	
Fluctuating	
Unsurveyed *Indefinite*	
Man-made	
LAKES *Label as required* Perennial *When too numerous to show individual lakes, show representative pattern and descriptive note.*	numerous small lakes 756 618
Non-Perennial *(dry, intermittent, etc.) Illustration includes small perennial lake*	
RESERVOIRS Natural Shorelines	
Man-made Shorelines *Label when necessary for clarity*	reservoir
Too small to show to scale	reservoir
Under Construction	under construction

VFR AERONAUTICAL CHARTS - Topographic Information

HYDROGRAPHY	
STREAMS	
Perennial	
Non-Perennial	
Fanned Out *Alluvial fan*	
Braided	
Disappearing	
Seasonally Fluctuating *with undefined limits*	
with maximum bank limits, prominent and constant	
Sand Deposits In and Along Riverbeds	

HYDROGRAPHY	
WET SAND AREAS *Within and adjacent to desert areas*	
AQUEDUCTS	aqueduct
Abandoned or Under Construction	abandoned aqueduct
Underground	underground aqueduct
Suspended or Elevated	
Tunnels	
Kanats *Underground aqueduct with air vents*	underground aqueduct

HYDROGRAPHY	
FLUMES, PENSTOCKS AND SIMILAR FEATURES	flume
Elevated	flume
Underground	underground flume
FALLS Double-Line	falls
Single-Line	falls
RAPIDS Double-Line	rapids
Single-Line	rapids
CANALS	ERIE

HYDROGRAPHY	
To Scale	
Abandoned or Under Construction	abandoned
Abandoned to Scale	abandoned
SMALL CANALS AND DRAINAGE / IRRIGATION DITCHES Perennial	
Non-Perennial	
Abandoned or Ancient	
Numerous *Representative pattern and/or descriptive note.*	
Numerous	numerous canals and ditches

VFR AERONAUTICAL CHARTS - Topographic Information

HYDROGRAPHY	
SALT EVAPORATORS AND SALT PANS MAN EXPLOITED	
SWAMPS, MARSHES AND BOGS	
HUMMOCKS AND RIDGES	
MANGROVE AND NIPA	
PEAT BOGS	peat bog
TUNDRA	tundra
CRANBERRY BOGS	cranberry bog
RICE PADDIES *Extensive areas indicated by label only.*	

HYDROGRAPHY	
LAND SUBJECT TO INUNDATION	
SPRINGS, WELLS AND WATERHOLES	
GLACIERS	
GLACIAL MORAINES	
ICE CLIFFS	
SNOWFIELDS, ICE FIELDS AND ICE CAPS	9000 8000 7000
ICE PEAKS	
FORESHORE FLATS *Tidal flats exposed at low tide.*	

VFR AERONAUTICAL CHARTS - Topographic Information

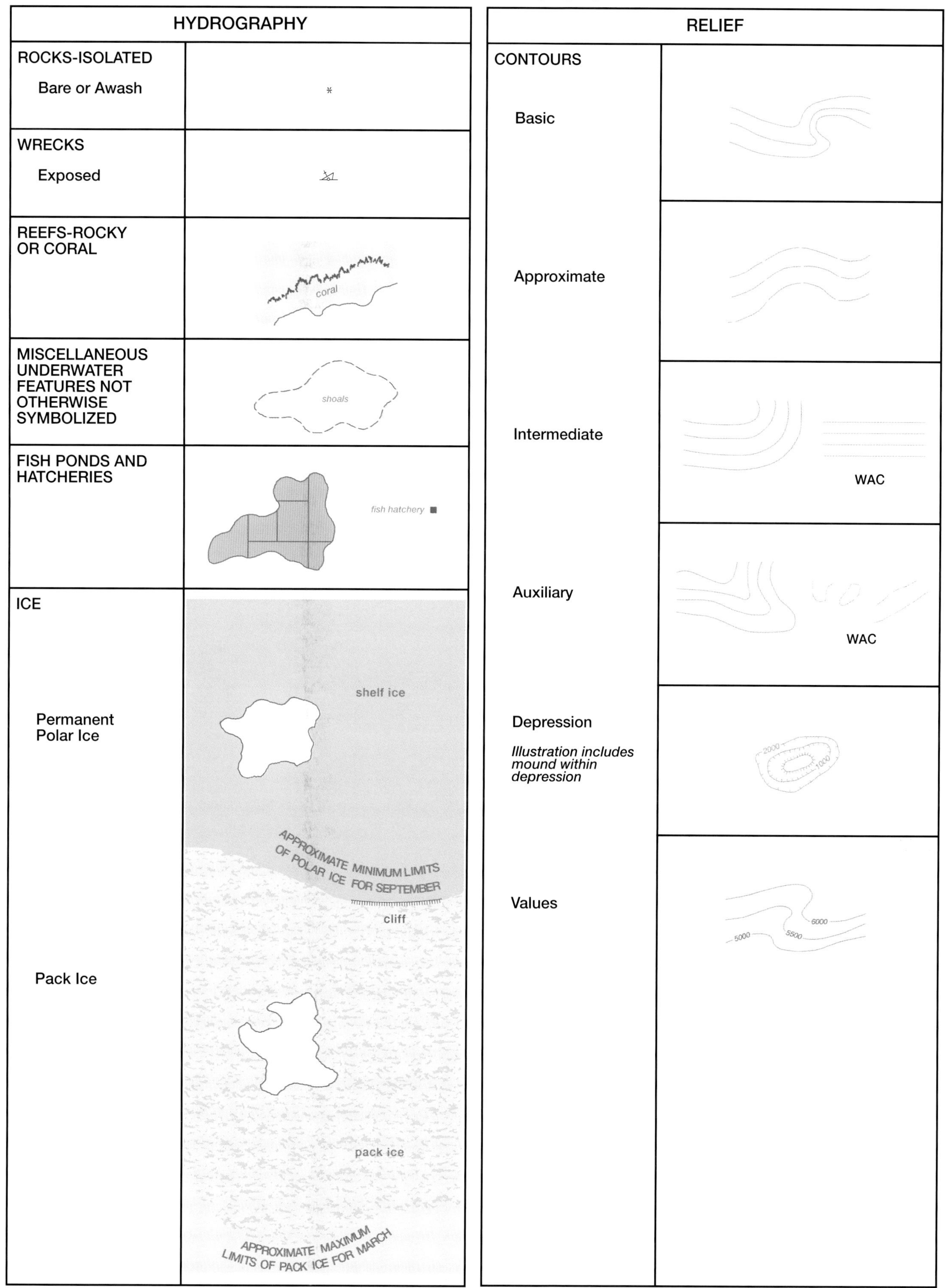

RELIEF	
SPOT ELEVATIONS Position Accurate	2216
Position Accurate, Elevation Approximate	2260
Approximate location	2119
Highest in General Area	6973 6973 WAC
Highest on Chart	12770
MOUNTAIN PASS	12632
HACHURING	
UNSURVEYED AREAS *Label appropriately as required*	UNSURVEYED
UNCONTOURED AREAS *Label appropriately as required*	RELIEF DATA INCOMPLETE
DISTORTED SURFACE AREAS	lava
LAVA FLOWS	

RELIEF	
SAND OR GRAVEL AREAS	
SAND RIDGES To Scale	
SAND DUNES To Scale	
SHADED RELIEF	
ROCK STRATA OUTCROP	rock strata
QUARRIES TO SCALE	quarry
STRIP MINES, MINE DUMPS AND TAILINGS To Scale	strip mine mine dump
CRATERS	crater crater
ESCARPMENTS, BLUFFS, CLIFFS, DEPRESSIONS, ETC.	
LEVEES AND ESKERS	levee

AIRPORTS

LANDPLANE

All recognizable runways, including some which may be closed, are shown for visual identification.

Public

Private

HELIPORT

Heliports public and private

Hospital Helipads

Trauma Center

Helipads located at major airports

SEAPLANE

ULTRALIGHT FLIGHT PARK

AIRPORT DATA GROUPING

Boxed airport name indicates airport for which a Special Traffic Rule has been established.

(Pvt) - Non-public use having emergency or landmark value.

Rotating Beacon in operation Sunset to Sunrise

FSS
NO SVFR
NAME (NAM) (PNAM)
CT -119.1 ★ Ⓒ (119.8 HELI)
ATIS 115.4
ASOS/ AWOS 135.42
03 L 122.95
AOE

FSS	Flight Service Station on field
NO SVFR	Airspace where fixed wing special visual flight rules operations are prohibited (shown above airport name) F.A.R. 91.
[box]	Indicates F.A.R. 93 Special Air Traffic Rules and Airport Traffic
(NAM)	Location Identifier
(PNAM)	ICAO Location Indicator
CT - 119.1	Control Tower (CT) - primary frequency
★	Star indicates operation part-time. See tower frequencies tabulation for hours of operation.
ATIS 115.4	Automatic Terminal Information Service
ASOS / AWOS 135.42	Automated Surface Weather Observing Systems (shown where full-time ATIS is not available). Some ASOS/AWOS facilities may not be located at airports.
03	Elevation in feet
L	Lighting in operation Sunset to Sunrise
*L	Lighting limitations exist, refer to Airport/Facility Directory.
122.95	UNICOM - Aeronautical advisory station
Ⓒ	Indicates Common Traffic Advisory Frequencies (CTAF)
(Unverified)	Unverified Heliport
AOE	Airport of Entry

When information is lacking, the respective character is replaced by a dash. Lighting codes refer to runway edge lights and may not represent the longest runway or full length lighting. Dashes are not shown on heliports or helipads unless additional information follows the elevation (e.g. UNICOM, CTAF).

RADIO AIDS TO NAVIGATION

VHF OMNI-DIRECTIONAL RADIO (VOR) RANGE

VOR-DME
PROVO
108.4 Ch 21 PVU

33 0 3 6 30 27 24 21 18 15 12

Open circle symbol shown when NAVAID located on airport. Type of NAVAID shown in top of box.

Compass Rose is "reference" oriented to magnetic north.

VOR

Operates less than continuous or On-Request

Transcribed Weather Broadcast (TWEB)

AMEDEE
*109.0 Ch 27 AHC

Underline indicates no voice on this frequency.

VORTAC

When an NDB NAVAID shares the same name and Morse Code as the VOR NAVAID the frequency can be colocated inside the same box to conserve space.

NDB Frequency — Name — ASOS/AWOS

PONTIAC
379 111.0 Ch 47 PTK

Morse Code

Frequency — Channel — Identifier

VOR-DME

Hazardous Inflight Weather Advisory Service (HIWAS)

SALEM
114.3 Ch 90 SVM

Crosshatch indicates Shutdown status

NON-DIRECTIONAL RADIOBEACON (NDB)

MONTAGUE
382 MOG

Underline indicates no voice on this frequency.

NDB-DME

GAMBELL
369 GAM
DME Ch 92 (114.5)

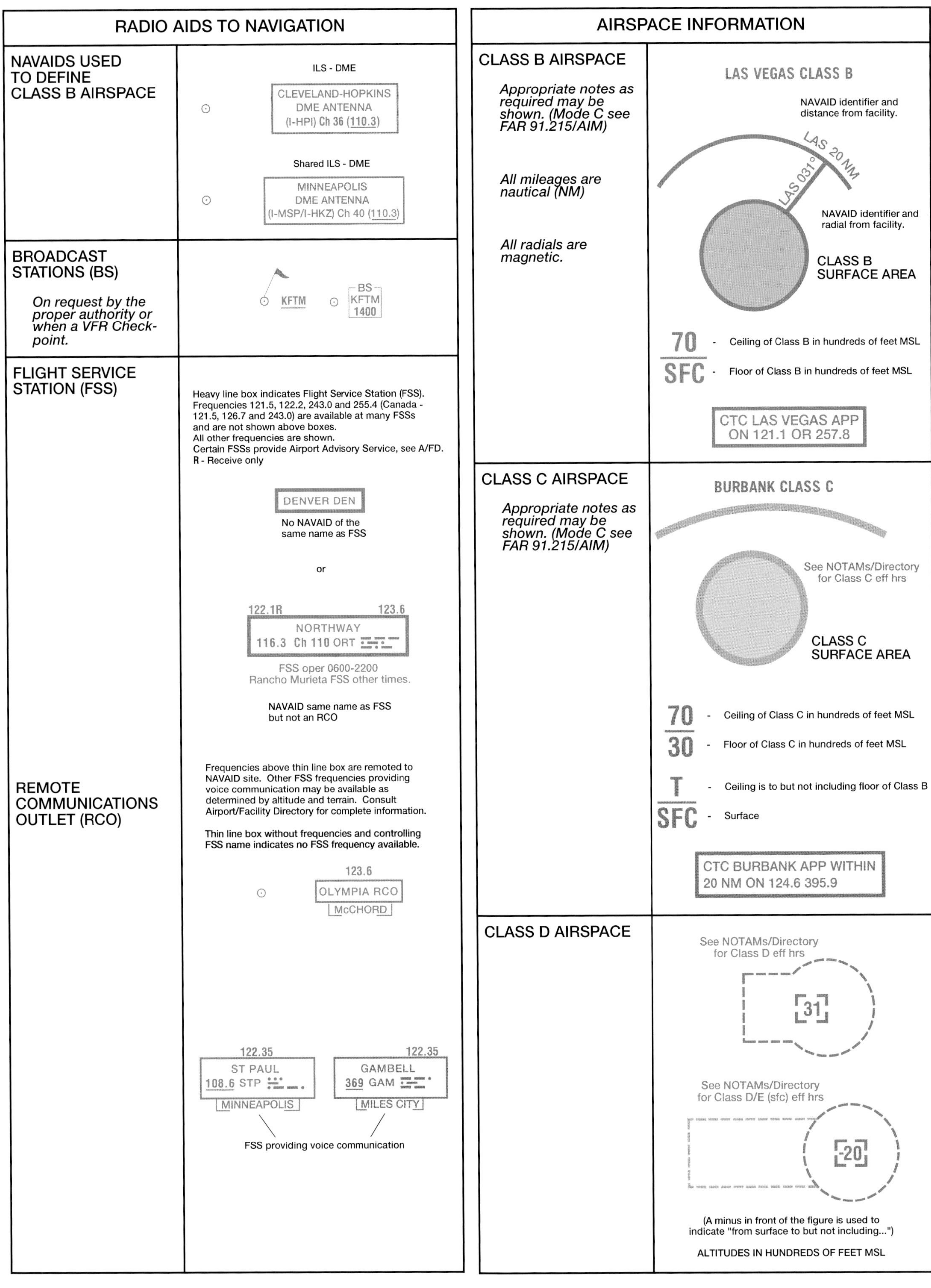

RADIO AIDS TO NAVIGATION
NAVAIDS USED TO DEFINE CLASS B AIRSPACE
ILS - DME
CLEVELAND-HOPKINS
DME ANTENNA
(I-HPI) Ch 36 (110.3)
Shared ILS - DME
MINNEAPOLIS
DME ANTENNA
(I-MSP/I-HKZ) Ch 40 (110.3)
BROADCAST STATIONS (BS)
On request by the proper authority or when a VFR Check-point.
KFTM
BS
KFTM
1400
FLIGHT SERVICE STATION (FSS)
Heavy line box indicates Flight Service Station (FSS).
Frequencies 121.5, 122.2, 243.0 and 255.4 (Canada - 121.5, 126.7 and 243.0) are available at many FSSs and are not shown above boxes.
All other frequencies are shown.
Certain FSSs provide Airport Advisory Service, see A/FD.
R - Receive only
DENVER DEN
No NAVAID of the same name as FSS
or
122.1R
123.6
NORTHWAY
116.3 Ch 110 ORT
FSS oper 0600-2200
Rancho Murieta FSS other times.
NAVAID same name as FSS but not an RCO
REMOTE COMMUNICATIONS OUTLET (RCO)
Frequencies above thin line box are remoted to NAVAID site. Other FSS frequencies providing voice communication may be available as determined by altitude and terrain. Consult Airport/Facility Directory for complete information.
Thin line box without frequencies and controlling FSS name indicates no FSS frequency available.
123.6
OLYMPIA RCO
McCHORD
122.35
ST PAUL
108.6 STP
MINNEAPOLIS
122.35
GAMBELL
369 GAM
MILES CITY
FSS providing voice communication
AIRSPACE INFORMATION
CLASS B AIRSPACE
Appropriate notes as required may be shown. (Mode C see FAR 91.215/AIM)
All mileages are nautical (NM)
All radials are magnetic.
LAS VEGAS CLASS B
NAVAID identifier and distance from facility.
LAS 20 NM
LAS 031°
NAVAID identifier and radial from facility.
CLASS B SURFACE AREA
70 - Ceiling of Class B in hundreds of feet MSL
SFC - Floor of Class B in hundreds of feet MSL
CTC LAS VEGAS APP ON 121.1 OR 257.8
CLASS C AIRSPACE
Appropriate notes as required may be shown. (Mode C see FAR 91.215/AIM)
BURBANK CLASS C
See NOTAMs/Directory for Class C eff hrs
CLASS C SURFACE AREA
70 - Ceiling of Class C in hundreds of feet MSL
30 - Floor of Class C in hundreds of feet MSL
T - Ceiling is to but not including floor of Class B
SFC - Surface
CTC BURBANK APP WITHIN 20 NM ON 124.6 395.9
CLASS D AIRSPACE
See NOTAMs/Directory for Class D eff hrs
31
See NOTAMs/Directory for Class D/E (sfc) eff hrs
-20
(A minus in front of the figure is used to indicate "from surface to but not including...")
ALTITUDES IN HUNDREDS OF FEET MSL

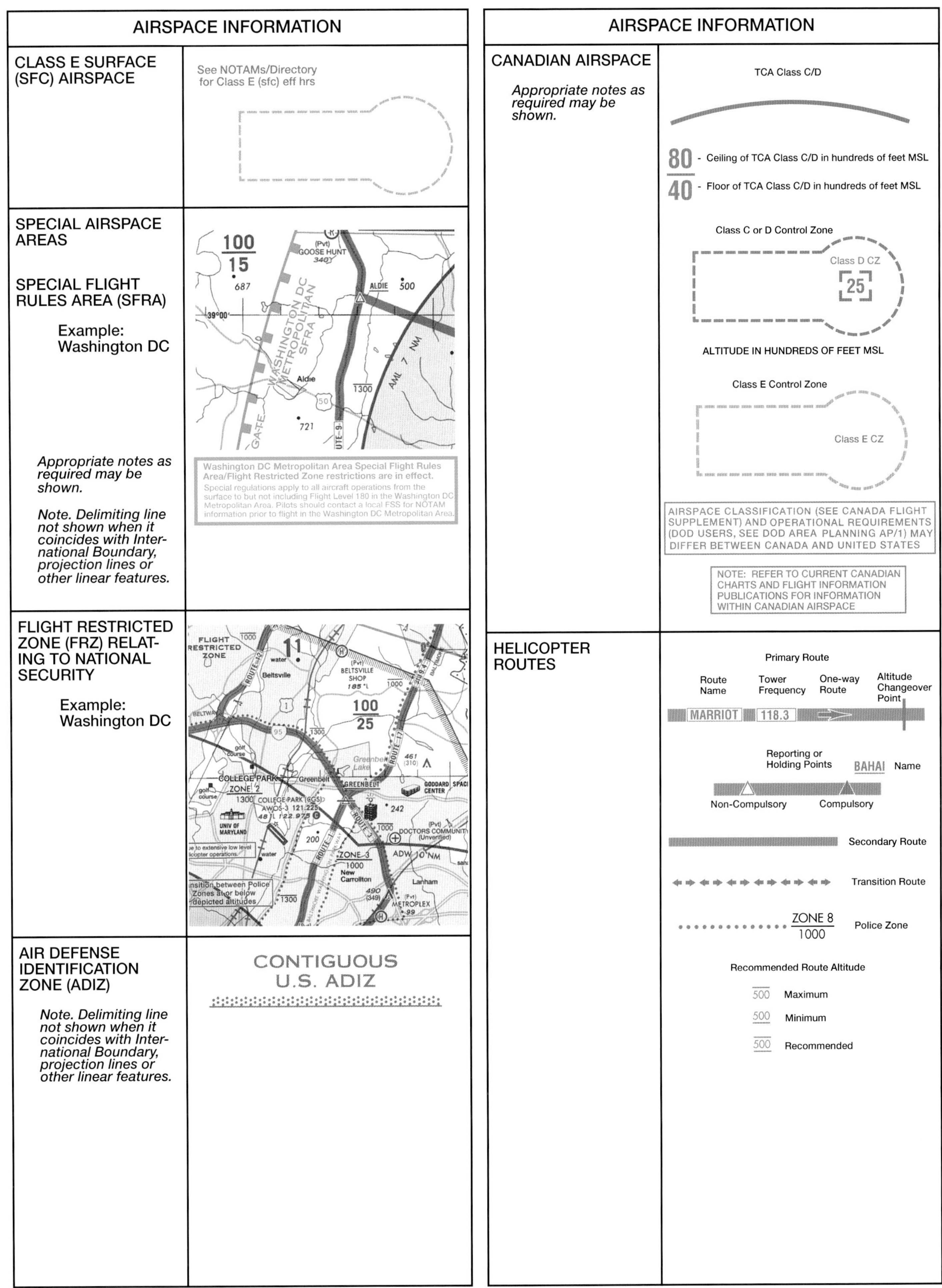

AIRSPACE INFORMATION
CLASS E SURFACE (SFC) AIRSPACE
See NOTAMs/Directory for Class E (sfc) eff hrs
SPECIAL AIRSPACE AREAS
SPECIAL FLIGHT RULES AREA (SFRA)
Example: Washington DC
Appropriate notes as required may be shown.
Note. Delimiting line not shown when it coincides with International Boundary, projection lines or other linear features.
100
15
(Pvt) GOOSE HUNT 340
687
ALDIE 500
39°00'
WASHINGTON DC METROPOLITAN SFRA
Aldie
1300
AML 7 NM
50
721
Washington DC Metropolitan Area Special Flight Rules Area/Flight Restricted Zone restrictions are in effect.
Special regulations apply to all aircraft operations from the surface to but not including Flight Level 180 in the Washington DC Metropolitan Area. Pilots should contact a local FSS for NOTAM information prior to flight in the Washington DC Metropolitan Area.
FLIGHT RESTRICTED ZONE (FRZ) RELATING TO NATIONAL SECURITY
Example: Washington DC
FLIGHT RESTRICTED ZONE
1000
11
water
Beltsville
(Pvt) BELTSVILLE SHOP 185
100
25
1300
COLLEGE PARK
ZONE 2
Greenbelt
GREENBELT
GODDARD SPACE CENTER
COLLEGE PARK (CGS) AWOS-3 121.225
UNIV OF MARYLAND
242
200
ZONE 3
1000
New Carrollton
DOCTORS COMMUNITY (Unverified)
ADW 10 NM
Lanham
Transition between Police Zones at or below depicted altitudes
METROPLEX
AIR DEFENSE IDENTIFICATION ZONE (ADIZ)
Note. Delimiting line not shown when it coincides with International Boundary, projection lines or other linear features.
CONTIGUOUS U.S. ADIZ
AIRSPACE INFORMATION
CANADIAN AIRSPACE
Appropriate notes as required may be shown.
TCA Class C/D
80 - Ceiling of TCA Class C/D in hundreds of feet MSL
40 - Floor of TCA Class C/D in hundreds of feet MSL
Class C or D Control Zone
Class D CZ
25
ALTITUDE IN HUNDREDS OF FEET MSL
Class E Control Zone
Class E CZ
AIRSPACE CLASSIFICATION (SEE CANADA FLIGHT SUPPLEMENT) AND OPERATIONAL REQUIREMENTS (DOD USERS, SEE DOD AREA PLANNING AP/1) MAY DIFFER BETWEEN CANADA AND UNITED STATES
NOTE: REFER TO CURRENT CANADIAN CHARTS AND FLIGHT INFORMATION PUBLICATIONS FOR INFORMATION WITHIN CANADIAN AIRSPACE
HELICOPTER ROUTES
Primary Route
Route Name
Tower Frequency
One-way Route
Altitude Changeover Point
MARRIOT
118.3
Reporting or Holding Points
BAHAI
Name
Non-Compulsory
Compulsory
Secondary Route
Transition Route
ZONE 8
1000
Police Zone
Recommended Route Altitude
500 Maximum
500 Minimum
500 Recommended

HELICOPTER ROUTE CHARTS - Aeronautical Information

AIRSPACE INFORMATION

SPECIAL USE AIRSPACE *Only the airspace effective below 18,000 feet MSL is shown.* *The type of area shall be spelled out in large areas if space permits.*	P-56 or R-6401 or W-518 PROHIBITED, RESTRICTED or WARNING AREA FALCON 1 MOA or A-631 MILITARY OPERATIONS AREA (MOA) or ALERT AREA
MILITARY TRAINING ROUTES (MTR)	VR269
SPECIAL AIR TRAFFIC RULES / AIRPORT TRAFFIC AREAS (FAR PART 93) *Appropriate boxed notes as required shown adjacent to area.*	SPECIAL NOTICE Pilots are required to obtain an ATC clearance prior to entering this area.
MODE C (FAR 91.215) *Appropriate notes as required may be shown.*	MODE C 30 NM
MISCELLANEOUS AIRSPACE AREAS Parachute Jumping Area with Frequency Glider Operating Area Ultralight Activity Hang Glider Activity	122.9 G U H
SPECIAL CONSERVATION AREAS National Park, Wildlife Refuge, Primitive and Wilderness Areas, etc.	HAVASU LAKE NATIONAL WILDLIFE REFUGE
TERMINAL RADAR SERVICE AREA (TRSA) *Appropriate notes as required may be shown.*	PALM SPRINGS TRSA SEE TWR FREQ TAB 80 - Ceiling of TRSA in hundreds of feet MSL 40 - Floor of TRSA in hundreds of feet MSL

NAVIGATIONAL AND PROCEDURAL INFORMATION

VFR CHECKPOINTS	Pictorial STATE CAPITOL STACKS 122.2 FRANCIS PEAK RCO CEDAR CITY (Pvt) LEWIS 420 H
VFR WAYPOINTS Stand-Alone Colocated with Visual Checkpoint Colocated with Visual Checkpoint & Reporting Point	VPXYZ NAME (VPXYZ) NAME (VPXYZ)
OBSTRUCTIONS *High-intensity lights may operate part-time or by proximity activation.*	bldg 1000′ AGL and higher 300′ AGL and higher or Group Obstruction or Obstruction with high-intensity lights. 2049 Elevation of the top above mean sea level (1149) Height above ground UC Under Construction or reported and position / elevation unverified
MAXIMUM ELEVATION FIGURE (MEF) *(see page 2 for explanation).*	124
NAVIGATION DATA	N38°56.32′ W76°36.91′ POWER PLANT N32°27.12′ W70°15.73′ ATL 25 NM ATL 033/25 NM N33°59.18′ W84°10.62′ ATL 033°

HELICOPTER ROUTE CHARTS - Topographic Information

NAVIGATIONAL AND PROCEDURAL INFORMATION

WARNING AND CAUTION NOTES	WARNING Extensive fleet and air operations being conducted in offshore areas to approximately 100 miles seaward. CAUTION: Be prepared for loss of horizontal reference at low altitude over lake during hazy conditions and at night.
LOCAL MAGNETIC NOTES Unreliability Notes	Magnetic disturbance of as much as 78° exists at ground level and 10° or more at 3000 feet above ground level in this vicinity.

CULTURE

RAILROADS Single Track Double Track	
ROADS Dual-Lane: Divided Highways Major Boulevards & Major Streets Primary	HOLLYWOOD BOULEVARD 495 95 25
BRIDGES	Railroad Road
POPULATED PLACES Built-up Areas	
BOUNDARIES International State or Province	

CULTURE

POWER TRANSMISSION LINES	
PROMINENT PICTORIALS	TEMPLE
LANDMARKS	Landmark-stadium, factory, school, etc. Lookout Tower Mines or Quarries Race Track Outdoor Theater Tank-water, oil or gas

HYDROGRAPHY

SHORELINES	
MAJOR LAKES AND RIVERS	
RESERVOIRS	Dam

RELIEF

SPOT ELEVATIONS Position Accurate	*405*

VFR FLYWAY PLANNING CHARTS - Aeronautical Information

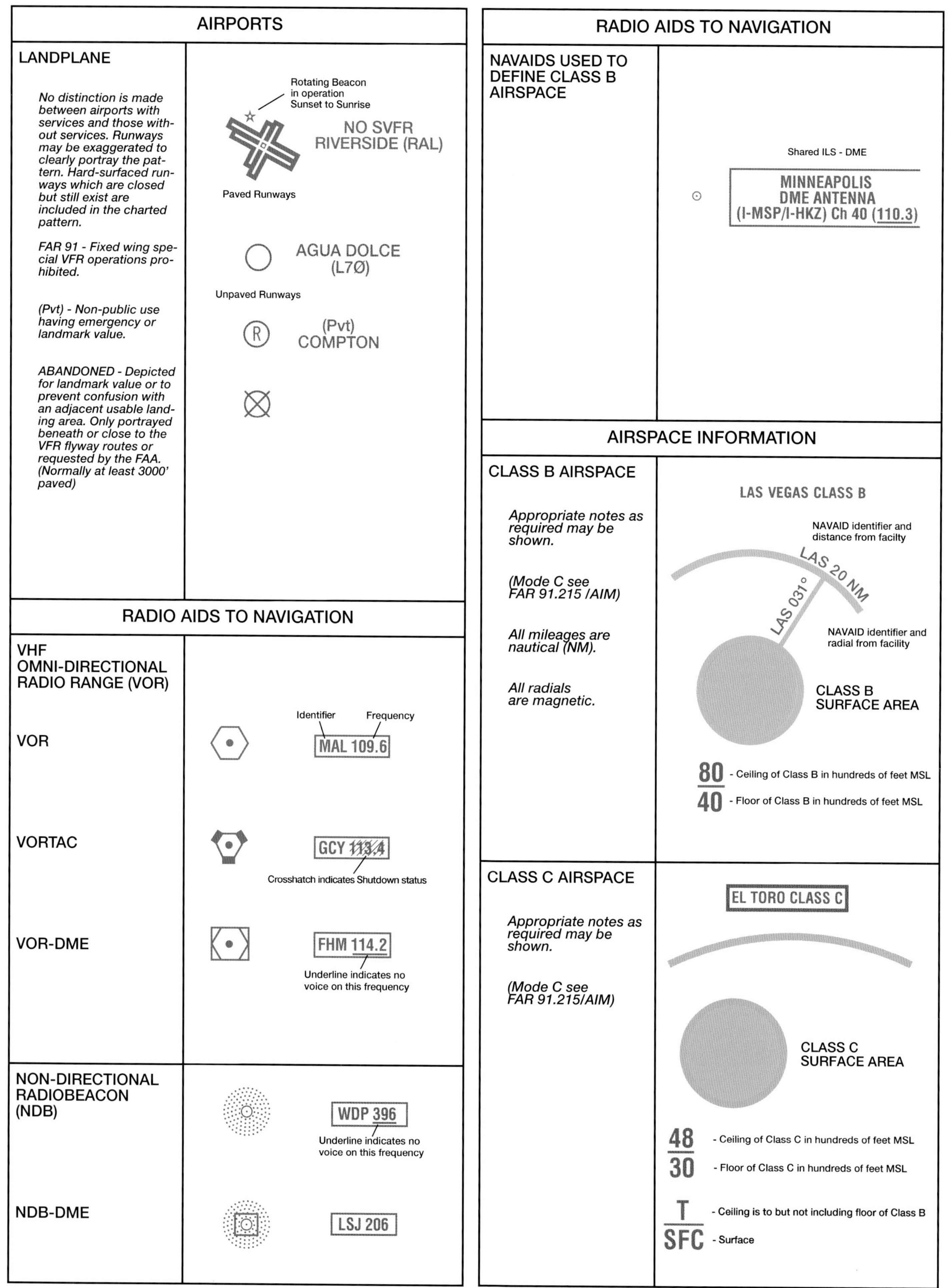

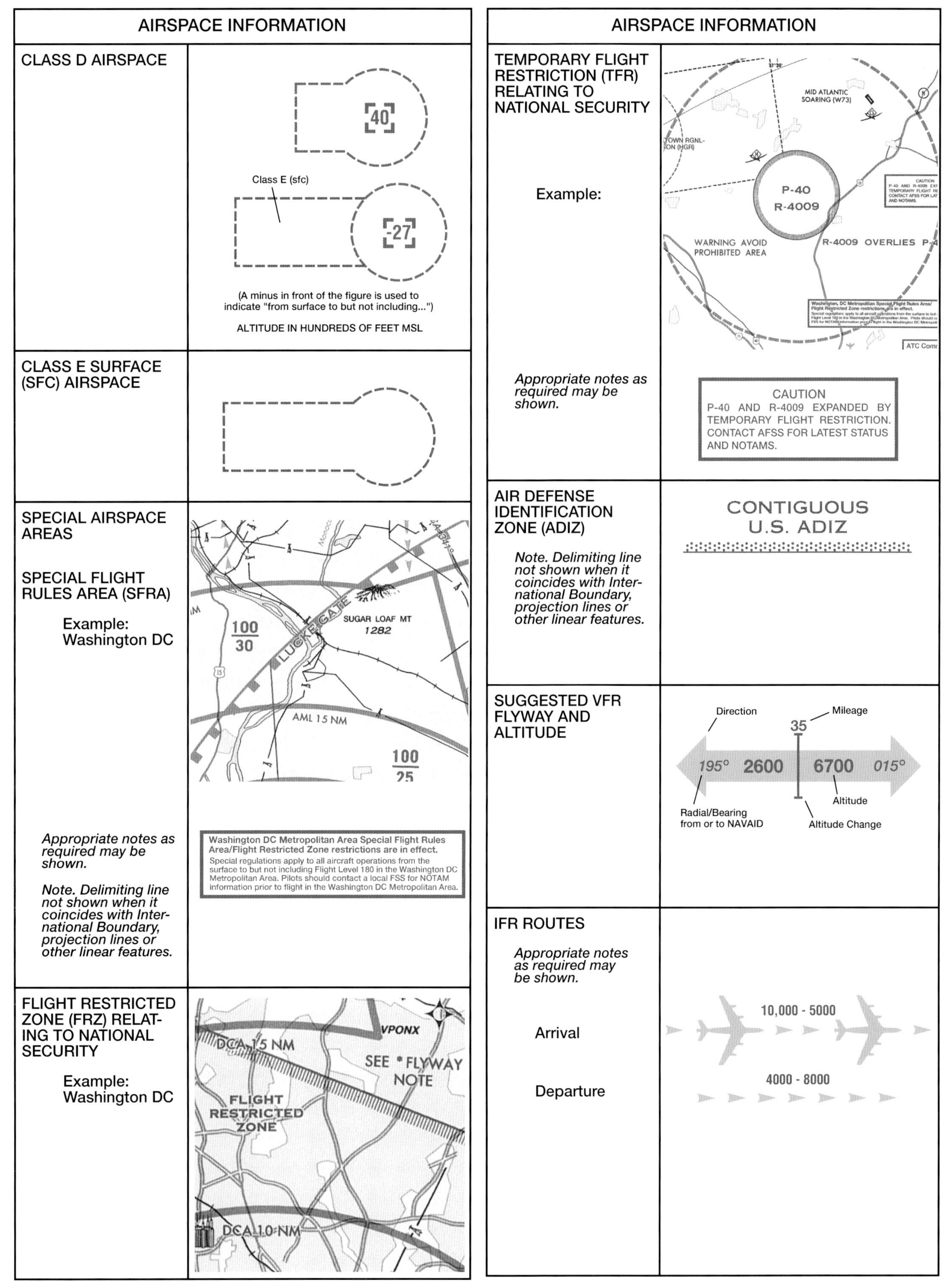
AIRSPACE INFORMATION
CLASS D AIRSPACE
40
Class E (sfc)
-27
(A minus in front of the figure is used to indicate "from surface to but not including...")
ALTITUDE IN HUNDREDS OF FEET MSL
CLASS E SURFACE (SFC) AIRSPACE
SPECIAL AIRSPACE AREAS
SPECIAL FLIGHT RULES AREA (SFRA)
Example: Washington DC
SUGAR LOAF MT
1282
100
30
LUCKETGATE
AML 15 NM
100
25
Appropriate notes as required may be shown.
Washington DC Metropolitan Area Special Flight Rules Area/Flight Restricted Zone restrictions are in effect.
Special regulations apply to all aircraft operations from the surface to but not including Flight Level 180 in the Washington DC Metropolitan Area. Pilots should contact a local FSS for NOTAM information prior to flight in the Washington DC Metropolitan Area.
Note. Delimiting line not shown when it coincides with International Boundary, projection lines or other linear features.
FLIGHT RESTRICTED ZONE (FRZ) RELATING TO NATIONAL SECURITY
Example: Washington DC
VPONX
DCA 15 NM
SEE * FLYWAY NOTE
FLIGHT RESTRICTED ZONE
DCA 10 NM
AIRSPACE INFORMATION
TEMPORARY FLIGHT RESTRICTION (TFR) RELATING TO NATIONAL SECURITY
Example:
MID ATLANTIC SOARING (W73)
P-40
R-4009
WARNING AVOID PROHIBITED AREA
R-4009 OVERLIES P-4
Appropriate notes as required may be shown.
CAUTION
P-40 AND R-4009 EXPANDED BY TEMPORARY FLIGHT RESTRICTION. CONTACT AFSS FOR LATEST STATUS AND NOTAMS.
AIR DEFENSE IDENTIFICATION ZONE (ADIZ)
Note. Delimiting line not shown when it coincides with International Boundary, projection lines or other linear features.
CONTIGUOUS U.S. ADIZ
SUGGESTED VFR FLYWAY AND ALTITUDE
Direction
Mileage
35
195° 2600
6700 015°
Radial/Bearing from or to NAVAID
Altitude
Altitude Change
IFR ROUTES
Appropriate notes as required may be shown.
Arrival
10,000 - 5000
Departure
4000 - 8000

AIRSPACE INFORMATION

TRANSITION ROUTES *Appropriate notes as required may be shown.*	VFR TRANSITION ROUTE ATC CLEARANCE REQUIRED SEE SHOWBOAT GRAPHIC ON SIDE PANEL
Uni-directional	
Bi-directional	
SPECIAL USE AIRSPACE *Only the airspace effective below 18,000 feet MSL is shown.* *The type of area shall be spelled out in large areas if space permits.*	P-56 or R-6401 or W-518 PROHIBITED, RESTRICTED or WARNING AREA FALCON 1 MOA or A-631 MILITARY OPERATIONS AREA (MOA) or ALERT AREA
MILITARY TRAINING ROUTES (MTR)	IR21
SPECIAL AIR TRAFFIC RULES / AIRPORT TRAFFIC AREAS (FAR Part 93) *Appropriate boxed note as required shown adjacent to area.*	
MODE C (FAR 91.215) *Appropriate notes as required may be shown.*	MODE C 30 NM
TERMINAL RADAR SERVICE AREA (TRSA)	PALM SPRINGS TRSA 100 - Ceiling of TRSA in hundreds of feet MSL 90 - Floor of TRSA in hundreds of feet MSL

AIRSPACE INFORMATION

MISCELLANEOUS AIRSPACE AREAS	
Parachute Jumping Area	
Glider Operating Area	G
Ultralight Activity	U
Hang Glider Activity	H

NAVIGATIONAL AND PROCEDURAL INFORMATION

VFR CHECKPOINTS	LA PORTE Pictorial STADIUM HARVEY (S43) NORTHBROOK
VFR WAYPOINTS Stand-Alone	*VPXYZ*
Collocated with Visual Checkpoint	NAME *(VPXYZ)*
OBSTRUCTIONS *Only those obstacles specified by the local ATC Facility shall be shown.* *Above Ground Level (AGL) heights are not shown.* *High-intensity lights may operate part-time or by proximity activation.*	Pictorial 352 629 808 less than 1000′ AGL 2562 5612 1000′ AGL and higher 2049 922 974 4920 Group Obstruction High-intensity Lights

NAVIGATIONAL AND PROCEDURAL INFORMATION	
NAVIGATIONAL DATA	N38°56.32′ W76°36.91′
	POWER PLANT N32°27.12′ W70°15.73′
	ATL 25 NM; ATL 033/25 NM N33°59.18′ W84°10.62′; ATL 033°
CULTURE	
RAILROADS — Single and Multiple Tracks	
ROADS — Dual-Lane Divided Highway	HARBOR FREEWAY 110
Primary	40
POPULATED PLACES — Built-up Areas	BREMERTON
Towns	LAWRENCEVILLE
BOUNDARIES — International	
POWER TRANSMISSION LINES	
PROMINENT PICTORIALS	TEMPLE
LANDMARKS	POWER PLANT

HYDROGRAPHY	
SHORELINES	
MAJOR LAKES AND RIVERS	Bridge
RESERVOIRS	Dam
RELIEF	
Spot Elevations — Position Accurate Mountain Peaks	6504

IFR AERONAUTICAL CHARTS

EXPLANATION OF IFR ENROUTE TERMS AND SYMBOLS

The discussions and examples in this section will be based primarily on the IFR (Instrument Flight Rule) Enroute Low Altitude Charts. Other IFR products use similar symbols in various colors (see Section 3 of this guide). The chart legends list aeronautical symbols with a brief description of what each symbol depicts. This section will provide a more detailed discussion of some of the symbols and how they are used on IFR charts.

FAA charts are prepared in accordance with specifications of the Interagency Air Cartographic Committee (IACC), and are approved by representatives of the Federal Aviation Administration and the Department of Defense. Some information on these charts may only apply to military pilots.

AIRPORTS

Active airports with hard-surfaced runways of 3000' or longer are shown on FAA IFR Low Altitude Enroute Charts for the contiguous United States. Airports with hard or soft runways of 3000' or longer are shown on IFR Low Altitude Alaska Charts. Airports with runways of 5000' or longer are shown on IFR High Altitude Enroute Charts. Airports with hard or soft runways of 4000' or longer are shown on IFR High Altitude Alaska Enroute Charts. Active airports with approved instrument approach procedures are also shown regardless of runway length or composition.

Charted airports are classified according to the following criteria:

Blue – Facilities with an approved Department of Defense (DoD) Low Altitude Instrument Approach Procedure and/or DoD RADAR MINIMA published in the DOD FLIP (Flight Information Publication or the FAA U.S. Terminal Procedures Publication (TPP).
Green – Facilities with an approved Low Altitude Instrument Approach Procedure published in the FAA TPP volumes.
Brown – Facilities without a published Instrument Approach Procedure or RADAR MINIMA.

Airports are plotted in their true geographic position unless the symbol conflicts with a radio aid to navigation (NAVAID) at the same location. In such cases, the airport symbol may be displaced, but the relationship between the airport and the NAVAID is retained.

Airports are identified by the airport name. In the case of military airports, the abbreviated letters AFB (Air Force Base), NAS (Naval Air Station), NAF (Naval Air Facility), MCAS (Marine Corps Air Station), AAF (Army Air Field), etc., appear as part of the airport name.

Airports marked "Pvt" immediately following the airport name are not for public use, but otherwise meet the criteria for charting as specified above.

Runway length is the length of the longest active runway (including displaced thresholds but excluding overruns) and is shown to the nearest 100 feet using 70 feet as the division point; e.g., a runway of 8,070' is labeled 81.

The following runway compositions (materials) constitute a hard-surfaced runway: asphalt, bitumen, concrete, and tar macadam. Runways that are not hard-surfaced have a small letter "s" following the runway length, indicating a soft surface.

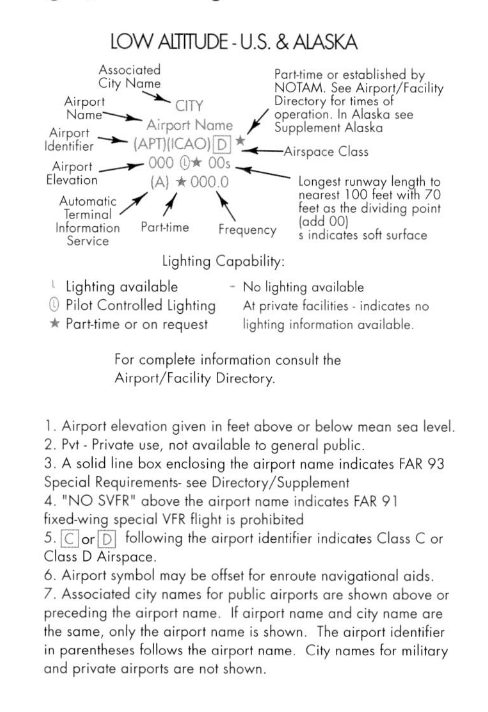

A symbol following the elevation under the airport name means that runway lights are in operation sunset to sunrise. A symbol indicates there is Pilot Controlled Lighting. A symbol means the lighting is part-time or on request. The pilot should consult the Airport/Facility Directory for light operating procedures. The Aeronautical Information Manual thoroughly explains the types and uses of airport lighting aids.

RADIO AIDS TO NAVIGATION (NAVAIDs)

All IFR radio NAVAIDs that have been flightchecked and are operational are shown on IFR enroute charts. VHF/UHF NAVAIDs (VORs, TACANs, and UHF NDBs) are shown in black, and LF/MF NAVAIDs (Compass Locators and Aeronautical or Marine NDBs) are shown in brown.

On enroute charts, information about NAVAIDs is boxed as illustrated below. To avoid duplication of data, when two or more NAVAIDs in a general area have the same name, the name is usually printed only once inside an identification box with the frequencies, TACAN channel numbers, identification letters, or Morse Code identifications of the different NAVAIDs all shown in appropriate colors.

NAVAIDS that have a status of shutdown will have the frequency and channel number crosshatched. Use of the NAVAID status "Shutdown" is only used when a facility has been decommissioned but cannot be published as such because of pending airspace actions.

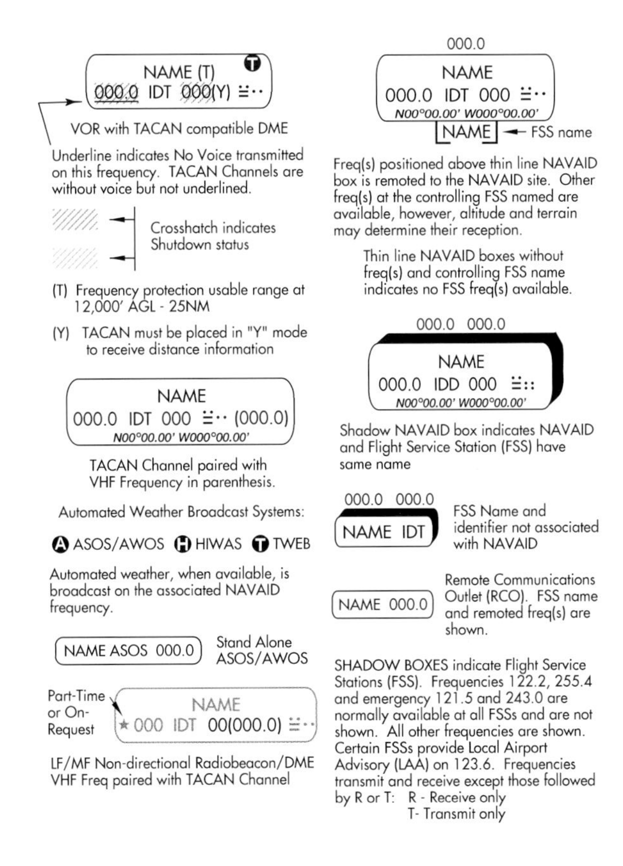

CONTROLLED AIRSPACE

Controlled airspace consists of those areas where some or all aircraft may be subjected to air traffic control within the following airspace classifications of A, B, C, D, & E.

Class A Airspace is depicted as open area (white) on the Enroute High Charts. It consists of airspace from 18,000 MSL to FL600.

Class B Airspace is depicted as screened blue area with a solid line encompassing the area.

Class C Airspace is depicted as screened blue area with a dashed line encompassing the area with a [C] following the airport name.

Class B and Class C Airspace consist of controlled airspace extending upward from the surface or a designated floor to specified altitudes, within which all aircraft and pilots are subject to the operating rules and requirements specified in the Federal Aviation Regulations (FAR) 71. Class B and C Airspace are shown in abbreviated forms on Enroute Low Altitude charts. A general note adjacent to Class B airspace refers the user to the appropriate VFR Terminal Area Chart.

Class D Airspace (airports with an operating control tower) are depicted as open area (white) with a [D] following the airport name.

Class E Airspace is depicted as open area (white) on the Enroute Low Charts. It consists of airspace below 18,000 MSL.

UNCONTROLLED AIRSPACE

Class G Airspace within the United States extends to 14,000' MSL. This uncontrolled airspace is shown as screened brown.

Air Route Traffic Control Centers (ARTCC) are established to provide Air Traffic Control to aircraft operating on IFR flight plans within controlled airspace, particularly during the enroute phase of flight. Boundaries of the ARTCCs are shown in their entirety using the symbol below. Center names are shown adjacent and parallel to the boundary line.

NEW YORK
WASHINGTON
Air Route Traffic Control Center (ARTCC)

ARTCC sector frequencies are shown in boxes outlined by the same symbol.

WASHINGTON
Hagerstown
134.15 385.4
ARTCC Remoted Sites with discrete VHF and UHF frequencies

SPECIAL USE AIRSPACE

Special use airspace confines certain flight activities or restricts entry, or cautions other aircraft operating within specific boundaries. Special use airspace areas are depicted on aeronautical charts. Special use airspace areas are shown in their entirety, even when they overlap, adjoin, or when an area is designated within another area. The areas are identified by type and identifying number or name (R-4001), effective altitudes, operating time, weather conditions (VFR/IFR) during which the area is in operation, and voice call of the controlling agency, on the back or front panels of the chart. Special Use Airspace with a floor of 18,000' MSL or above is not shown on the Enroute Low Altitude Charts. Similarly, Special Use Airspace with a ceiling below 18,000' MSL is not shown on Enroute High Altitude Charts.

SPECIAL USE AIRSPACE

P-56
R-123
W-789
CYA-101
CYD-102
CYR-103

A-456

WALL 1 MOA
WALL 2 MOA
EXCLUSION AREA
Line delimits internal separation of same Special Use Area

P - Prohibited Area
W - Warning Area
R - Restricted Area

LOW ALTITUDE ONLY
MOA - Military Operations Area
A - Alert Area

Canada Only:
CYA - Advisory Area
CYD - Danger Area
CYR - Restricted Area

Caribbean Only:
D - Danger Area

SEE AIRSPACE TABULATION FOR COMPLETE INFORMATION

OTHER AIRSPACE

FAR 93 Special Airspace Traffic Rules are shown with a solid line box around the airport name.

Mode C Required Airspace (from the surface to 10,000' MSL) within 30 NM radius of the primary airport(s) for which a Class B airspace is designated, is depicted on Enroute Low Altitude Charts. Mode C is also depicted within 10 NM of all airports listed in Appendix D of FAR 91.215 and the Aeronautical Information Manual (AIM).

Mode C is also required for operations within and above all Class C airspace up to 10,000' MSL, but not depicted. See FAR 91.215 and the Aeronautical Information Manual(AIM).

Airports within which fixed-wing special VFR flight is prohibited are shown as:

INSTRUMENT AIRWAYS

The FAA has established two fixed route systems for air navigation. The VOR and LF/MF (low or medium frequency) system—designated from 1,200' AGL to but not including 18,000' MSL—is shown on Low Altitude Enroute Charts, and the Jet Route system—designated from 18,000' MSL to FL 450 inclusive—is shown on High Altitude Enroute Charts.

VOR LF/MF AIRWAY SYSTEM (LOW ALTITUDE ENROUTE CHARTS)

In this system VOR airways—airways based on VOR or VORTAC NAVAIDs—are depicted in black and identified by a "V" (Victor) followed by the route number (e.g., "V12"). In Alaska and Canada, some segments of low-altitude airways are based on LF/MF navaids and are charted in brown instead of black.

LF/MF airways—airways based on LF/MF NAVAIDs—are sometimes called "colored airways" because they are identified by color name and number (e.g., "Amber One", charted as "A1"). Green and Red airways are plotted east and west, and Amber and Blue airways are plotted north and south. Regardless of their color identifier, LF/MF airways are shown in brown. U.S. colored airways exist only in Alaska, those within the conterminous U.S. have been rescinded.

AIRWAY/ROUTE DATA

On both series of Enroute Charts, airway/route data such as the airway identifications, bearings or radials, mileages, and altitude (e.g., MEA, MOCA, MAA) are shown aligned with the airway and in the same color as the airway.

Airways/Routes predicated on VOR or VORTAC NAVAIDs are defined by the outbound radial from the NAVAID. Airways/Routes predicated on LF/MF NAVAIDs are defined by the inbound bearing.

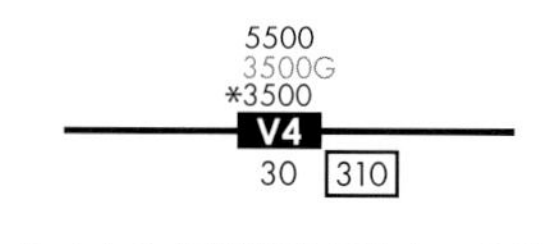

Victor Route (with RNAV/GPS MEA shown in blue)

AREA NAVIGATION (RNAV) "T" ROUTE SYSTEM

The FAA has created new low altitude area navigation (RNAV) routes for the en route and terminal environments. The RNAV routes will provide more direct routing for IFR aircraft and enhance the safety and efficiency of the National Airspace System. To utilize these routes aircraft will need to be equipped with IFR approved Global Navigation Satellite System (GNSS). In Alaska, TSO-145a and 146a equipment is required. Low altitude RNAV only routes are identified by the letter "T" prefix, followed by a three digit number (T-200 to T-500). Routes are depicted in aeronautical blue on the IFR Enroute Low Altitude charts. RNAV route data (route line, identification boxes, mileages, waypoints, waypoint names, magnetic reference bearings, and MEAs) will also be printed in aeronautical blue. Magnetic reference bearings will be shown originating from a waypoint, fix/reporting point or NAVAID. A GNSS minimum IFR en route altitude (MEA) for each segment will be established to ensure obstacle clearance and communications reception. MEAs will be identified with a "G" suffix.

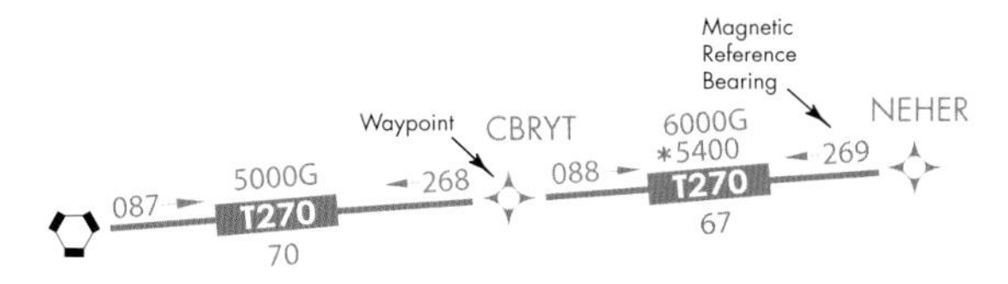

Joint Victor/RNAV routes will be charted as outlined above except as noted. The joint Victor route and the RNAV route identification box shall be shown adjacent to each other. Magnetic reference bearings will not be shown. MEAs will be stacked in pairs or in two separate columns, GNSS and Victor. On joint routes, RNAV specific information will be printed in blue.

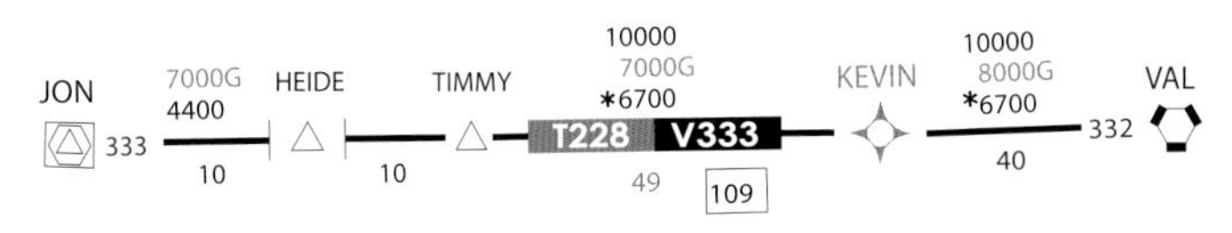

OFF ROUTE OBSTRUCTION CLEARANCE ALTITUDE (OROCA)

The Off Route Obstruction Clearance Altitude (OROCA) is represented in thousands and hundreds of feet above mean sea level. The OROCA represents the highest possible elevation including both terrain and other vertical obstructions (towers, trees., etc.) bounded by the ticked lines of latitude and longitude. In this example the OROCA represents 12,500 feet.

12 5

OROCA is computed just as the Maximum Elevation Figure (MEF) found on Visual charts except that it provides an additional vertical buffer of 1,000 feet in designated non-mountainous areas and a 2,000 foot vertical buffer in designated mountainous areas within the United States. For areas in Mexico and the Caribbean, located outside the U.S. ADIZ, the OROCA provides obstruction clearance with a 3,000 foot vertical buffer. Unlike a MEF, when determining an OROCA the area 4 NM around each quadrant is analyzed for obstructions. Evaluating the area around the quadrant provides the chart user the same lateral clearance an airway provides should the line of intended flight follow a ticked line of latitude or longitude. OROCA does not provide for NAVAID signal coverage, communication coverage and would not be consistent with altitudes assigned by Air Traffic Control. OROCAs can be found over all land masses and open water areas containing man-made obstructions (such as oil rigs). OROCAs are shown in every 30 x 30 minute quadrant on Area Charts, every one degree by one degree quadrant for U.S. Low Altitude Enroute Charts and every two degree by two degree quadrant on Alaska Low Enroute Charts.

MILITARY TRAINING ROUTES (MTRs)

Military Training Routes (MTRs) are routes established for the conduct of low-altitude, highspeed military flight training (generally below 10,000 feet MSL at airspeeds in excess of 250 knots IAS). These routes are depicted in brown on Enroute Low Altitude Charts, and are not shown on inset charts or on IFR Enroute High Altitude Charts. Enroute Low Altitude Charts depict all IR (IFR Military Training Route) and VR (VFR Military Training Route) routes, except those VRs that are entirely at or below 1500 feet AGL.

Military Training Routes are identified by designators (IR-107, VR-134) which are shown in brown on the route centerline. Arrows indicate the direction of flight along the route. The width of the route determines the width of the line that is plotted on the chart: Route segments with a width of 5 NM or less, both sides of the centerline, are shown by a .02" line. IR-000

Route segments with a width greater than 5 NM, either or both sides of the centerline, are shown by a .035" line. VR-000

MTRs for particular chart pairs (ex. L1/2, etc.) are alphabetically, then numerically tabulated. The tabulation is located on the title panel and includes MTR type and unique ident and altitude range.

JET ROUTE SYSTEM (HIGH ALTITUDE ENROUTE CHARTS)

Jet routes are based on VOR or VORTAC navaids, and are depicted in black with a "J" identifier followed by the route number (e.g., "J12"). In Alaska, Russia and Canada some segments of jet routes are based on LF/MF navaids and are shown in brown instead of black.

AREA NAVIGATION (RNAV) "Q" ROUTE SYSTEM (HIGH ALTITUDE ENROUTE CHARTS)

The FAA has adopted certain amendments to Title 14, Code of Federal Regulations which paved the way for the development of new area navigation (RNAV) routes in the U.S. National Airspace System (NAS). These amendments enable the FAA to take advantage of technological advancements in navigation systems such as the Global Positioning System (GPS). RNAV "Q" Route MEAs are shown when other than 18,000'. MEAs for GNSS RNAV aircraft are identified with a "G" suffix. MEAs for DME/DME/IRU RNAV aircraft do not have a "G" suffix. RNAV routes and associated data

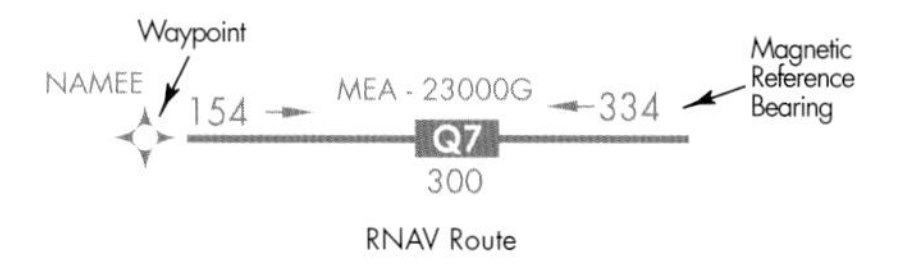

RNAV Route

are charted in aeronautical blue. Magnetic reference bearings are shown originating from a waypoint, fix/reporting point, or NAVAID. Joint Jet/RNAV route identification boxes will be located adjacent to each other with the route charted in black. With the exception of Q-Routes in the Gulf of Mexico, GNSS or DME/DME/IRU RNAV are required, unless otherwise indicated. Radar monitoring is required. DME/DME/IRU RNAV aircraft should refer to the A/FD for DME information. Altitude values are stacked highest to lowest.

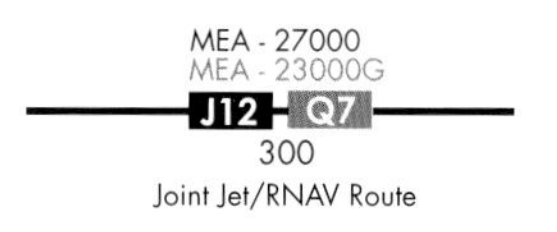

Joint Jet/RNAV Route

TERRAIN CONTOURS ON AREA CHARTS

Based on a recommendation of the National Transportation Safety Board, terrain has been added to the Enroute Area Charts to increase pilots' situational awareness of terrain in the terminal area and to increase the safety of flight. The following Area Charts are affected: Anchorage, Denver, Detroit, Fairbanks, Juneau, Los Angeles, Phoenix, Prudhoe Bay, San Francisco, Vancouver and Washington.

When terrain rises at least a 1,000 feet above the primary airports' elevation, terrain is charted using shades of brown with brown contour lines and values. The initial contour will be 1,000 or 2,000 feet above the airports' elevation. Subsequent intervals will be 2,000 or 3,000 foot increments.

Contours are supplemented with a representative number of spots elevations and are shown in solid black. The highest elevation on an Area Chart is shown with a larger spot and text.

Any uncontrolled airspace boundaries will be depicted with a .012" brown line and a .060" screen brown band on the uncontrolled side.

The following boxed notes are added to affected Area Charts as necessary:

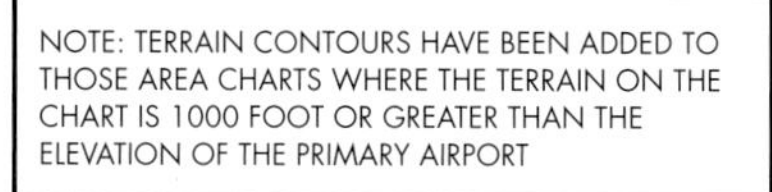
NOTE: TERRAIN CONTOURS HAVE BEEN ADDED TO THOSE AREA CHARTS WHERE THE TERRAIN ON THE CHART IS 1000 FOOT OR GREATER THAN THE ELEVATION OF THE PRIMARY AIRPORT

UNCONTROLLED AIRSPACE BOUNDARIES ARE DEPICTED WITH A SOLID BROWN LINE AND A .125" WIDE SHADED BROWN BAND. THE SHADED SIDE REPRESENTS THE UNCONTROLLED SIDE

IFR AERONAUTICAL CHART SYMBOLS

IFR Enroute Low/High Altitude (U.S. & Alaska Charts)

Oceanic Route Charts
North Atlantic and North Pacific Route Charts

GENERAL INFORMATION

Symbols shown are for the Instrument Flight Rules (IFR) Enroute Low and High Altitude Charts.

IFR ENROUTE LOW/HIGH ALTITUDE U.S. & ALASKA CHARTS

AIRPORTS

AIRPORT DATA	**LOW/HIGH ALTITUDE** Airports/Seaplane bases shown in BLUE and GREEN have an approved Instrument Approach Procedure published. Those in BLUE have an approved DoD Instrument Approach Procedure and/or DoD RADAR MINIMA published in DoD FLIPS or FAA TPP. Airports/Seaplane bases shown in BROWN do not have a published Instrument Approach Procedure. All IAP Airports are shown on the Low Altituide Charts. Non-IAP Airports shown on the U.S. Low Altitude Charts have a minimum hard surface runway of 3000'. Non-IAP Airports shown on the Alaska Low Altitude Charts have a minimum hard or soft surface runway of 3000'. Airports shown on the U.S. High Altitude Charts have a minimum hard surface runway of 5000'. Airports shown on the Alaska High Altitude Charts have a minimum hard or soft surface runway of 4000'. Associated city names for public airports are shown above or preceding the airport name. If airport name and city name are the same, only the airport name is shown. City names for military and private airports are not shown. The airport identifier in parentheses follows the airport name or Pvt. Airport symbol may be offset for enroute navigational aids. Pvt - Private Use
AIRPORT DATA DEPICTION	**LOW ALTITUDE - U.S. & ALASKA** Associated City Name → CITY Airport Name → Airport Name Airport Identifier → (APT)(ICAO) D ★ ← Airspace Class Part-time or established by NOTAM. See Airport/Facility Directory for times of operation. In Alaska see Supplement Alaska Airport Elevation → 000 Ⓛ★ 00s ← Longest runway length to nearest 100 feet with 70 feet as the dividing point (add 00) s indicates soft surface (A) ★ 000.0 — Automatic Terminal Information Service, Part-time, Frequency Lighting Capability: L Lighting available — No lighting available Ⓛ Pilot Controlled Lighting — At private facilities - indicates no lighting information available. ★ Part-time or on request For complete information consult the Airport/Facility Directory. 1. Airport elevation given in feet above or below mean sea level 2. Pvt - Private use, not available to general public. 3. A solid line box enclosing the airport name indicates FAR 93 Special Requirements- see Directory/Supplement 4. "NO SVFR" above the airport name indicates FAR 91 fixed-wing special VFR flight is prohibited 5. C or D following the airport identifier indicates Class C or Class D Airspace. 6. Airport symbol may be offset for enroute navigational aids. 7. Associated city names for public airports are shown above or preceding the airport name. If airport name and city name are the same, only the airport name is shown. The airport identifier in parentheses follows the airport name. City names for military and private airports are not shown.
	HIGH ALTITUDE - U.S. Associated City Name → CITY Airport Name → Airport Name (APT) ← Airport Identifier
	HIGH ALTITUDE - ALASKA Airport Name → Airport Name CITY ← Associated City Name (APT)(ICAO) ← Airport Identifier Airport Elevation → 000 00s ← Longest runway length to nearest 100 feet with 70 feet as the dividing point (add 00) s indicates soft surface (A) ★ 000.0 — Automatic Terminal Information Service, Part-time, Frequency

AIRPORTS

CIVIL	LOW/ HIGH ALTITUDE
CIVIL AND MILITARY	LOW/ HIGH ALTITUDE
MILITARY	LOW/ HIGH ALTITUDE
SEAPLANE - CIVIL	LOW ALTITUDE
HELIPORT	LOW ALTITUDE Ⓗ Ⓗ Ⓗ

RADIO AIDS TO NAVIGATION

VHF OMNIDIRECTIONAL RADIO RANGE (VOR)
DISTANCE MEASURING EQUIPMENT (DME)
TACTICAL AIR NAVIGATION (TACAN)

LOW/ HIGH ALTITUDE

VHF / UHF Data is depicted in Black
LF / MF Data is depicted in Brown

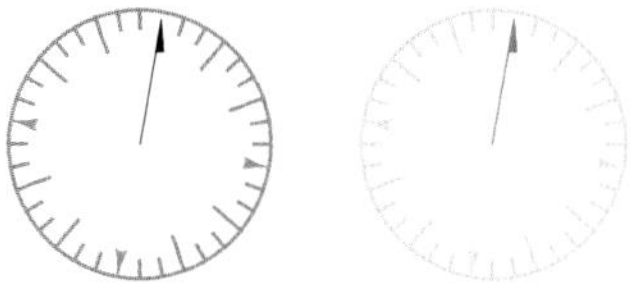

COMPASS ROSES are oriented to Magnetic North of the NAVAID which may not be adjusted to the charted isogonic values.

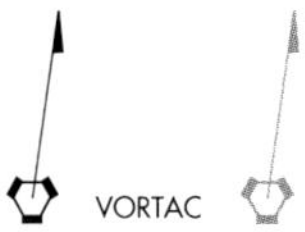

VORTAC

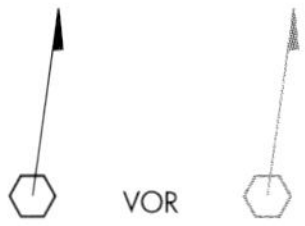

VOR

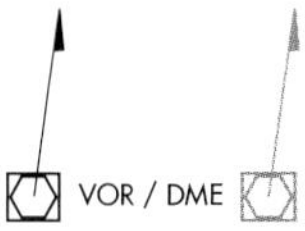

VOR / DME

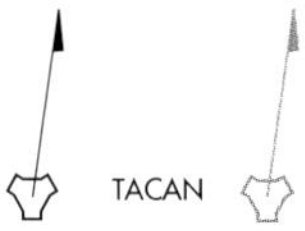

TACAN

"L" and "T" Category Radio Aids located off Jet Routes are depicted in screen black.

NON-DIRECTIONAL RADIO BEACON (NDB)
MARINE RADIO BEACON (RBN)

LOW/ HIGH ALTITUDE

NDB or RBN with Magnetic North Indicator

UHF NDB

NDB with DME

COMPASS LOCATOR BEACON

LOW ALTITUDE

ILS LOCALIZER

LOW ALTITUDE

ILS Localizer Course with additional navigation function.

VOR/DME RNAV WAYPOINT DATA

HIGH ALTITUDE - ALASKA

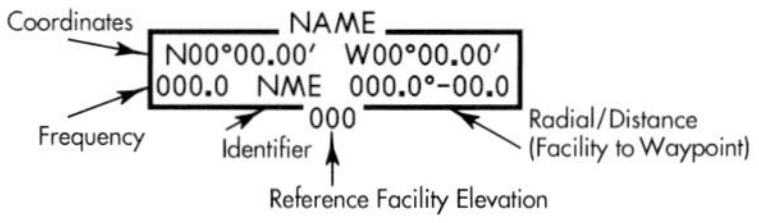

Coordinates — NAME — N00°00.00′ W00°00.00′ — 000.0 NME 000.0°-00.0 — 000

Frequency — Identifier — Reference Facility Elevation — Radial/Distance (Facility to Waypoint)

RNAV WAYPOINT

LOW/ HIGH ALTITUDE

NAMEE

RADIO AIDS TO NAVIGATION

NAVIGATION and COMMUNICATION BOXES

LOW/ HIGH ALTITUDE

PINE BLUFF (T) H
116.0 PBF 107(Y)
N34°14.81′ W91°55.57′

VOR with TACAN compatible DME

Underline indicates No Voice Transmitted on this frequency

TACAN channels are without voice but not underlined

Crosshatch indicates Shutdown status

(T) Frequency Protection - usable range 25 NM at 12000′ AGL

(Y) TACAN must be placed in "Y" mode to receive distance information

A ASOS/AWOS - Automated Surface Observing Station/Automated Weather Observing Station

H HIWAS - Hazardous Inflight Weather Advisory Service

T TWEB - Transcribed Weather Broadcast

Automated weather, when available, is broadcast on the associated NAVAID frequency.

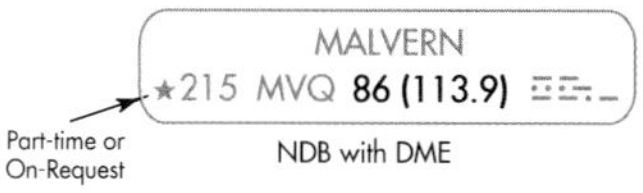

MALVERN
★215 MVQ 86 (113.9)

Part-time or On-Request

NDB with DME

DME channel and paired VHF frequency are shown

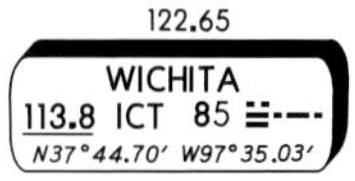

122.65
WICHITA
113.8 ICT 85
N37°44.70′ W97°35.03′

FSS associated with a NAVAID

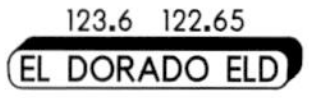

123.6 122.65
EL DORADO ELD

Name and identifier of FSS not associated with NAVAID

Shadow NAVAID Boxes indicate Flight Service Station (FSS) locations. Frequencies 122.2, 255.4 and emergency 121.5 and 243.0 are normally available at all FSSs and are not shown. All other frequencies are shown above the box.

Certain FSSs provide Local Airport Advisory (LAA) on 123.6.

Frequencies transmit and receive except those followed by R or T: R - Receive only T - Transmit only

In Canada, shadow boxes indicate FSSs with standard group frequencies of 121.5, 126.7 and 243.0.

JONESBORO 122.55

Remote Communications Outlet (RCO) FSS name and remoted frequency are shown

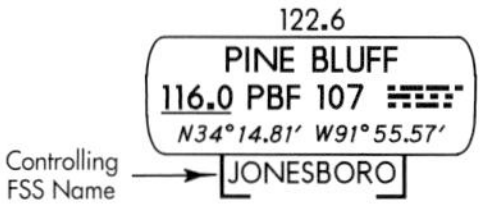

122.6
PINE BLUFF
116.0 PBF 107
N34°14.81′ W91°55.57′
Controlling FSS Name — JONESBORO

Thin Line NAVAID Boxes without frequencies and controlling FSS name indicate no FSS frequencies available. Frequencies positioned above thin line boxes are remoted to the NAVAID sites. Other frequencies at the controlling FSS named are available, however, altitude and terrain may determine their reception.

Morse Code is not shown in NAVAID boxes on High Altitude Charts.

⊙ Flight Service Station (FSS), Remote Communications Outlet (RCO) or Automated Weather Observing Station (AWOS/ASOS) not associated with a charted NAVAID or airport.

IFR ENROUTE LOW/HIGH ALTITUDE U.S. & ALASKA CHARTS

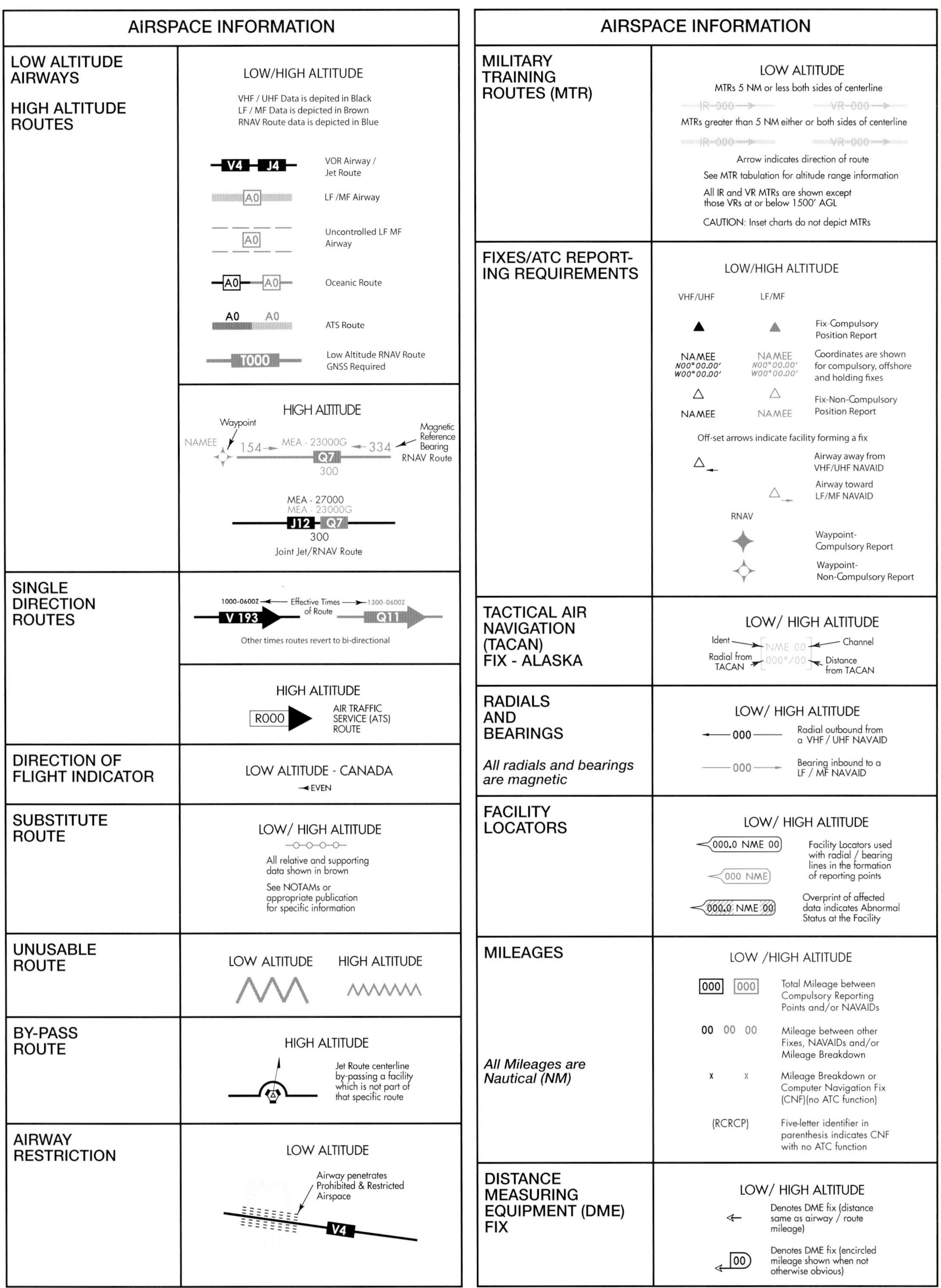

AIRSPACE INFORMATION

Item	Symbol / Description
MINIMUM ENROUTE ALTITUDE (MEA) *All Altitudes Are MSL Unless Otherwise Noted*	LOW ALTITUDE 3500 / 3000G V4 — RNAV/GPS MEA — 3500 A0 5500 → / ← 3500 V4 — Directional MEA — 5500 → / ← 3500 A0 HIGH ALTITUDE MEA-31000 J4 — Shown along Routes when other than 18,000′
MINIMUM ENROUTE ALTITUDE (MEA) GAP	LOW/HIGH ALTITUDE V4 MEA GAP — MEA is established when there is a gap in navigation signal coverage
MAXIMUM AUTHORIZED ALTITUDE (MAA) *All Altitudes Are MSL Unless Otherwise Noted*	LOW ALTITUDE MAA-15500 V4 — MAA-15500 A0 HIGH ALTITUDE MAA-41000 J4 — Shown along Routes when other than 45,000′
MINIMUM OBSTRUCTION CLEARANCE ALTITUDE (MOCA) *All Altitudes Are MSL Unless Otherwise Noted*	LOWALTITUDE 5500 / *3500 V4 ← MOCA → 5500 / *3500 A0 7000G / *6300 T266 112
CHANGEOVER POINT	LOW/ HIGH ALTITUDE 00 / 00 — VOR Changeover Point giving mileage to NAVAIDs (Not shown at midpoint locations)
ALTITUDE CHANGE	LOW/ HIGH ALTITUDE MEA, MOCA and / or MAA change at other than NAVAIDs
MINIMUM CROSSING ALTITUDE (MCA)	LOW/ HIGH ALTITUDE X NEHER V6 4000 SW — X DIGGS V6 4000 SW — X GRANT T244 7400 SE
MINIMUM RECEPTION ALTITUDE (MRA)	LOW/HIGH ALTITUDE R COPEL MRA 4500 — R SHIMY MRA 4500
HOLDING PATTERNS *RNAV Holding Pattern Magnetic Reference Bearing is determined by the isogonic value at the waypoint or fix.*	LOW/HIGH ALTITUDE NAMEE N00°00.00′ W00°00.00′ — V4 — NAMEE N00°00.00′ W00°00.00′ — Holding reporting points have coordinate values shown Left Turn — Right Turn (IAS) Holding Pattern with max. restricted airspeed 210K applies to altitudes above 6000′ to and including 14000′ 175K applies to all altitudes IAS: Indicated Airspeed Magnetic Reference Bearing → 245 — NAMEE — Waypoint — RNAV Holding

AIRSPACE INFORMATION

Item	Symbol / Description
AIR DEFENSE IDENTIFICATION ZONE (ADIZ)	LOW/ HIGH ALTITUDE CONTIGUOUS U.S. ADIZ ALASKA ADIZ / CANADA ADIZ — Adjoining ADIZ
AIR ROUTE TRAFFIC CONTROL CENTER (ARTCC)	LOW/ HIGH ALTITUDE NEW YORK / WASHINGTON WASHINGTON Hagerstown 134.15 385.4 — ARTCC Remoted Sites with discrete VHF and UHF frequencies
AIR TRAFFIC SERVICE IDENTIFICATION DATA	LOW/ HIGH ALTITUDE CTA/FIR MIAMI OCEANIC KZMA — Type of Area Traffic Service Ceiling Floor NY RADIO — Call Sign Frequency
ALTIMETER SETTING CHANGE	LOW ALTITUDE QNH ALTIMETER QNE
FLIGHT INFORMATION REGIONS (FIR)	LOW/ HIGH ALTITUDE MONTREAL FIR CZUL MONTREAL FIR CZUL / TORONTO FIR CZYZ — Adjoining FIR
CONTROL AREAS (CTA)	LOW/ HIGH ALTITUDE MIAMI OCEANIC CTA/FIR KZMA NEW YORK OCEANIC CTA/FIR KZNY / MIAMI OCEANIC CTA/FIR KZMA — Adjoining CTA
UPPER INFORMATION REGIONS (UIR) UPPER CONTROL AREAS (UTA)	HIGH ALTITUDE Adjoining UTA / UIR Adjoining FIR and UIR
ADDITIONAL CONTROL AREAS	LOW ALTITUDE CONTROL 1234L HIGH ALTITUDE CONTROL 1234H

AIRSPACE INFORMATION

OFF ROUTE OBSTRUCTION CLEARANCE ALTITUDE (OROCA)

LOW ALTITUDE

12 5

Example: 12,500 feet

OROCA is computed similarly to the Maximun Elevation Figure (MEF) found on Visual charts except that it provides an additional vertical buffer of 1,000 feet in designated non-mountainous areas and a 2,000 foot vertical buffer in designated mountainous areas within the United States.

SPECIAL USE AIRSPACE

LOW/ HIGH ALTITUDE

P-00
R-000
W-000
A-000
CYR-000
CYA-000
(MU) D-000

P - Prohibited Area
R - Restricted Area
W - Warning Area

Low Only
A - Alert Area

Canada Only
CYR - Restricted Area
CYD - Danger Area
CYA - Advisory Area

Caribbean Only
D - Danger Area

In the Caribbean, the first 2 letters represent the country code, i.e. MY: Bahamas, MU: Cuba

W-000A
W-000B

EXCLUSION AREA AND NOTE

Internal lines delimit separation of the same Special Use Areas or Exclusion Areas

SEE AIRSPACE TABULATION ON EACH CHART FOR COMPLETE INFORMATION ON:

AREA IDENTIFICATION
EFFECTIVE ALTITUDE
OPERATING TIME
CONTROLLING AGENCY VOICE CALL

SPECIAL USE AIRSPACE Continued

LOW ALTITUDE

MOA - Military Operations Area

WALL 1 MOA
WALL 2 MOA

EXCLUSION AREA AND NOTE

Internal lines delimit separation of the same Special Use Area or Exclusion Areas

SEE AIRSPACE TABULATION ON EACH CHART FOR COMPLETE INFORMATION ON:

AREA IDENTIFICATION
EFFECTIVE ALTITUDE
OPERATING TIME
CONTROLLING AGENCY VOICE CALL

AIRSPACE INFORMATION

CONTROLLED AIRSPACE

HIGH ALTITUDE

CLASS A AIRSPACE

Open Area (White)

That airspace from 18,000′ MSL to and including FL 600, including the airspace overlying the waters within 12 NM of the coast of the contiguous United States and Alaska and designated offshore areas, excluding Santa Barbara Island, Farallon Island, the airspace south of latitude 25 04′00″N, the Alaska peninsula west of longitude 160 00′00″W, and the airspace less than 1,500′ AGL.

That airspace from 18,000′ MSL to and including FL 450, including Santa Barbara Island, Farallon Island, the Alaska peninsula west of longitude 160 00′00″W, and designated offshore areas.

LOW ALTITUDE

CLASS B AIRSPACE

Screened Blue with a Solid Blue Outline

That airspace from the surface to 10,000′ MSL (unless otherwise designated) surrounding the nation's busiest airports. Each Class B airspace area is individually tailored and consists of a surface area and two or more layers.

MODE C AREA

A Solid Blue Outline

That airspace within 30 NM of the primary airports of Class B airspace and within 10 NM of designated airports. Mode-C transponder equipment is required. (see FAR 91.215)

CLASS B AIRSPACE SEE ATLANTA VFR TERMINAL AREA CHART FOR DETAILS

Mode C Area

LOW ALTITUDE

CLASS C AIRSPACE

Screened Blue with a Solid Blue Dashed Outline

That airspace from the surface to 4,000′ (unless otherwise designated) above the elevation of selected airports (charted in MSL). The normal radius of the outer limits of Class C airspace is 10 NM. Class C airspace is also indicated by the letter C in a box following the airport name.

LOW ALTITUDE

CLASS D AIRSPACE

Open Area (White)

That airspace, from the surface to 2,500′ (unless otherwise designated) above the airport elevation (charted in MSL), surrounding those airports that have an operational control tower. Class D airspace is indicated by the letter D in a box following the airport name.

IFR ENROUTE LOW/HIGH ALTITUDE U.S. & ALASKA CHARTS

AIRSPACE INFORMATION

CONTROLLED AIRSPACE	LOW ALTITUDE CLASS E AIRSPACE Open Area (White) That controlled airspace below 14,500′ MSL which is not Class B, C, or D. Federal airways from 1,200′ AGL to but not including 18,000′ MSL (unless otherwise specified). Other designated control areas below 14,500′ MSL. Not Charted That airspace from 14,500′ MSL to but not including 18,000′ MSL, including the airspace overlying the waters within 12 NM of the coast of the contiguous United States and Alaska and designated offshore areas, excluding the Alaska peninsula west of longitude 160 00′00″W and the airspace less than 1,500′ AGL.
CONTROLLED AIRSPACE Canada Only	LOW ALTITUDE CLASS B AIRSPACE Screened Brown Checkered Area Controlled airspace above 12,500′ MSL
UNCONTROLLED AIRSPACE	LOW/ HIGH ALTITUDE CLASS G AIRSPACE Screened Brown Area Low Altitude That portion of the airspace below 14,500′ MSL that has not been designated as Class B, C, D or E airspace. High Altitude That portion of the airspace from 18,000′ MSL and above that has not been designated as Class A airspace.
CANADIAN AIRSPACE *Appropriate notes as required may be shown.*	HIGH ALTITUDE DOD USERS REFER TO CURRENT DOD (NGA) CHARTS AND FLIGHT INFORMATION PUBLICATIONS FOR INFORMATION OUTSIDE OF U.S. AIRSPACE NOTE: REFER TO CURRENT CANADIAN CHARTS AND FLIGHT INFORMATION PUBLICATIONS FOR INFORMATION WITHIN CANADIAN AIRSPACE
AIRSPACE OUTSIDE OF U.S. *Other than Canada* *Appropriate notes as required may be shown.*	AIRSPACE CLASSIFICATION (SEE CANADA FLIGHT SUPPLEMENT) AND OPERATIONAL REQUIREMENTS (DOD USERS SEE DOD AREA PLANNING AP/1) MAY DIFFER BETWEEN CANADA AND THE UNITED STATES

NAVIGATIONAL AND PROCEDURAL INFORMATION

ISOGONIC LINE AND VALUE	LOW/ HIGH ALTITUDE 8°W Isogonic lines and values shall be based on the five year epoch.
TIME ZONE	LOW/ HIGH ALTITUDE Central Std +6=UTC ⁞ Eastern Std +5=UTC ‡ During periods of Daylight Saving Time (DT), effective hours will be one hour earlier than shown. All states observe DT except Arizona and Hawaii. ALL TIME IS COORDINATED UNIVERSAL TIME (UTC)
ENLARGEMENT AREA	LOW/ HIGH ALTITUDE JACKSONVILLE AREA CHART A-1
MATCH MARK	LOW/HIGH ALTITUDE

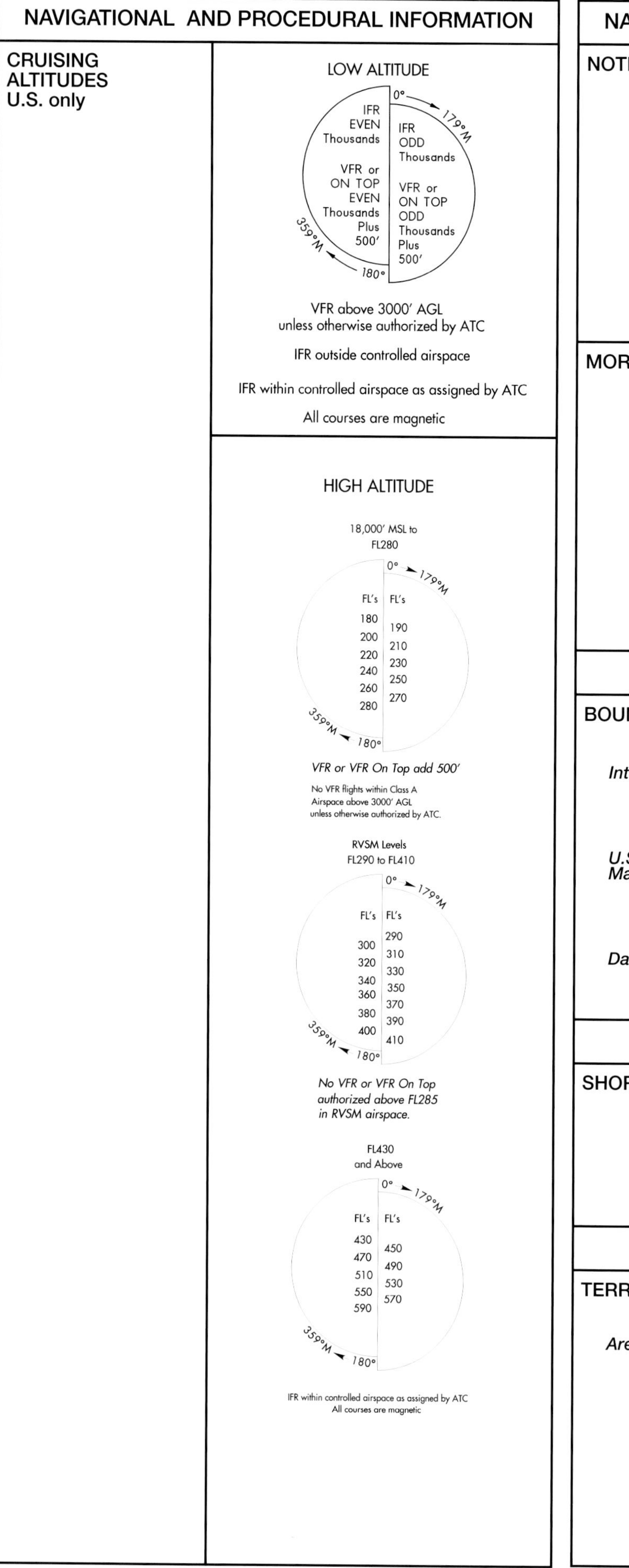

NAVIGATIONAL AND PROCEDURAL INFORMATION

NOTES

LOW/ HIGH ALTITUDE

FAA AIR TRAFFIC SERVICE OUTSIDE U.S. AIRSPACE IS PROVIDED IN ACCORDANCE WITH ARTICLE 12 AND ANNEX 11 OF ICAO CONVENTION. ICAO CONVENTION NOT APPLICABLE TO STATE AIRCRAFT BUT COMPLIANCE WITH ICAO STANDARDS AND PRACTICES IS ENCOURAGED.

CAUTION: POSSIBLE DAMAGE AND/OR INTERFERENCE TO AIRBORNE RADIO DUE TO HIGH LEVEL RADIO ENERGY IN THE VICINITY OF R-2206

CAUTION: ACCURACY OF AIR TRAFFIC SERVICES RELATIVE TO HAVANA FIR CANNOT BE CONFIRMED. CONSULT NOTAMS.

North American Datum of 1983 (NAD 83), for charting purposes is considered equivalent to World Geodetic System 1984 (WGS 84).

MORSE CODE

LOW/ HIGH ALTITUDE

A	·–	N	–·	1	·––––
B	–···	O	–––	2	··–––
C	–·–·	P	·––·	3	···––
D	–··	Q	––·–	4	····–
E	·	R	·–·	5	·····
F	··–·	S	···	6	–····
G	––·	T	–	7	––···
H	····	U	··–	8	–––··
I	··	V	···–	9	––––·
J	·–––	W	·––	0	–––––
K	–·–	X	–··–		
L	·–··	Y	–·––		
M	––	Z	––··		

CULTURE

BOUNDARIES

International

LOW/ HIGH ALTITUDE

Omitted when coincident with ARTCC or FIR

U.S. /Russia Maritime Line

LOW/ HIGH ALTITUDE

RUSSIA

UNITED STATES

Date Line

LOW/ HIGH ALTITUDE

INTERNATIONAL DATE LINE MONDAY SUNDAY

HYDROGRAPHY

SHORELINE

TOPOGRAPHY

TERRAIN

Area Charts

2000 2000 2000 4000 5600

AIRPORTS

AIRPORT DATA	Airport of Entry (AOE) are shown with four letter ICAO Identifier
LANDPLANE-CIVIL Refueling and repair facilities for normal traffic.	HONOLULU INTL (PHNL)
LANDPLANE-CIVIL AND MILITARY Refueling and repair facilities for normal traffic.	HILO INTL (PHTO)
LANDPLANE-MILITARY Refueling and repair facilities for normal traffic.	KALAELOA (PHJR)

RADIO AIDS TO NAVIGATION

		NARC	NPRC
VHF OMNIDIRECTIONAL RADIO RANGE (VOR) DISTANCE MEASURING EQUIPMENT (DME) TACTICAL AIR NAVIGATION (TACAN)	VOR		
	VOR / DME		
	VORTAC		
	TACAN		
NON-DIRECTIONAL RADIO BEACON (NDB) DISTANCE MEASURING EQUIPMENT (DME)	NDB		
	NDB / DME		

IDENTIFICATION BOX	Identification — MDY 400 — VHF Frequency N28°12.2′ W177°22.8′ — Latitude & Longitude Identification — NQM 347 — LF / MF Frequency CHAN 93 — TACAN Channel N28°12.2′ W177°22.8′ — Latitude & Longitude

AIRSPACE INFORMATION

AIR TRAFFIC SERVICE (ATS) OCEANIC ROUTES Note: Mileages are Nautical (NM)	A450 Identification 283 Mileage UB891 UHF Caribbean Identification 114 Mileage
ATS SINGLE DIRECTION ROUTE	A450
AERIAL REFUELING TRACKS	AR-900 E One Way FL 180/270 AR-903 E,W Two Way FL 180/270

AIRSPACE INFORMATION

AIR DEFENSE IDENTIFICATION ZONE (ADIZ)	HAWAIIAN ADIZ TAIWAN ADIZ JAPAN ADIZ
AIR ROUTE TRAFFIC CONTROL CENTER (ARTCC)	SEATTLE (ZSE) OAKLAND (ZOA)
FLIGHT INFORMATION REGIONS (FIR) and/or (CTA)	HONOLULU FIR PHZH HONIARA FIR ANAU HONOLULU FIR PHZH
UPPER INFORMATION REGIONS (UIR) UPPER CONTROL AREAS (UTA)	JAKARTA UIR WIIZ MERIDA UTA / UIR MMID MAZATLAN UTA / UIR MMZT MEXICO FIR / UIR MMFR FL 450
OCEANIC CONTROL AREAS (OCA) and /or (CTA /FIR)	OAKLAND OCEANIC CTA / FIR KZAK TOKYO FIR / OCA RJTG NAHA FIR / OCA RORG
ADDITIONAL OCEANIC CONTROL AREAS *Note: Limits not shown when coincident with Warning Areas.*	CONTROL 1485
BUFFER ZONE	Teeth point to area
NON-FREE FLYING ZONE	Teeth point to area
NORTH ATLANTIC / MINIMUM NAVIGATION PERFORMANCE SPECIFICATIONS (NAT/MNPS)	NAT MNPS (FL 285-FL420)
REPORTING POINTS	Name — ARTOP Latitude & Longitude — N20°52.7′ W80°00.0′ Compulsory Non-Compulsory Waypoint
SPECIAL USE AIRSPACE Warning Area	W-470 (NARC) W517 (NPRC)
Special Use	ATLANTIC FLEET WEAPONS RANGE
12 Mile Limit	
UNCONTROLLED AIRSPACE	

OCEANIC ROUTE CHARTS - Aeronautical Information

NAVIGATIONAL AND PROCEDURAL INFORMATION	
MILEAGE CIRCLES Note: Mileages are Nautical (NM)	100 NM
Time Zone Note: All time is Coordinated Universal (Standard) Time (UTC)	+3=UTC +2=UTC
Overlap Marks NPRC Only	S W
COMPASS ROSE *Note: Compass Roses oriented to Magnetic North*	MN 330 30 60 090 120 150 180 210 240 270 300
NOTES WARNING	WARNING AIRCRAFT INFRINGING UPON NON FREE FLYING TERRITORY MAY BE FIRED UPON WITHOUT WARNING WARNING UNLISTED RADIO EMISSIONS FROM THIS AREA MAY CONSTITUTE A NAVIGATION HAZARD OR RESULT IN BORDER OVERFLIGHT UNLESS UNUSUAL PRECAUTION IS EXERCISED.

CULTURAL BOUNDARIES	
INTERNATIONAL	
MARITIME	RUSSIA UNITED STATES
DATE LINE	MONDAY SUNDAY

HYDROGRAPHY	
SHORELINES	

EXPLANATION OF TPP TERMS AND SYMBOLS

The discussions and examples in this section will be based primarily on the IFR (Instrument Flight Rule) Terminal Procedures Publication (TPP). Other IFR products use similar symbols in various colors (see Section 2 of this guide). The publication legends list aeronautical symbols with a brief description of what each symbol depicts. This section will provide a more detailed discussion of some of the symbols and how they are used on TPP charts.

FAA charts are prepared in accordance with specifications of the Interagency Air Cartographic Committee (IACC), which are approved by representatives of the Federal Aviation Administration, and the Department of Defense. Some information on these charts may only apply to military pilots.

PILOT BRIEFING INFORMATION

The pilot briefing information format consists of three horizontal rows of boxed procedure-specific information along the top edge of the chart. Altitudes, frequencies and channel, course

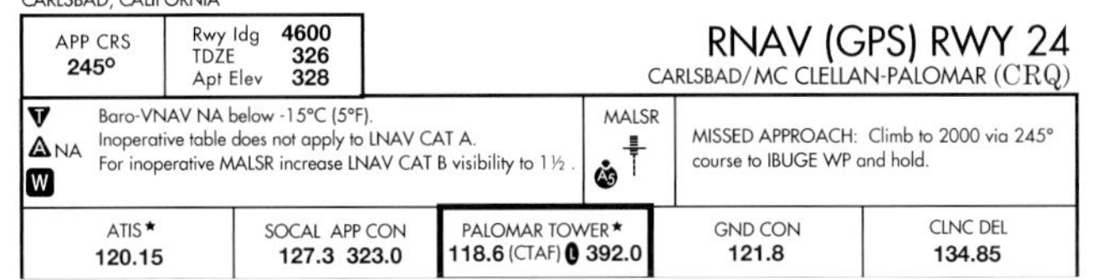

and elevation values (except HATs, HATHs and HAAs) are charted in bold type. The top row contains the primary procedure navigation information, final approach course, landing distance available, touchdown zone, threshold and airport elevations. The middle row contains procedure notes and limitations, icons indicating if nonstandard alternate and/or take-off minimums apply, approach lighting symbology, and the full text description of the missed approach procedure. The bottom row contains air to ground communication facilities and frequencies in the order in which they are used during an approach with the tower frequency box bolded.

NOTE: The W symbol indicates that outages of the WAAS vertical guidance may occur daily at this location due to initial system limitations. WAAS NOTAMs for vertical outages are not provided for this approach. Use LNAV minima for flight planning at these locations, whether as a destination or alternate. For flight operations at these locations, when the WAAS avionics indicate that LNAV/VNAV or LPV service is available, then vertical guidance may be used to complete the approach using the displayed level of service. Should an outage occur during the procedure, reversion to LNAV minima may be required. As the WAAS coverage is expanded, the W will be removed.

PLANVIEW

The data on the planview is drawn to scale, unless one of the following three charting devices are utilized: concentric rings, scale breaks or inset box(es). Most non-RNAV instrument procedure charts depict a reference or distance circle (not to be confused with the concentric rings) which is normally centered on the Final Approach Fix (FAF) and has a radius of 10NM. This circle is intended only to provide a sense of distance and scale. Data both within and without the circle is drawn to scale, unless a scale break symbol is utilized.

In most cases, obstructions close to the airport, which can be depicted within the parameters of the airport sketch, will be shown there rather than in the planview. Some of these obstacles may be controlling obstructions for the procedure.

Terrain Depiction

Terrain will be depicted in the planview portion of all IAPs at airports that meet the following criteria:

– If the terrain within the planview exceeds 4,000 feet above the airport elevation, or

– If the terrain within a 6.0 nautical mile radius of the Airport Reference Point (ARP) rises to at least 2,000 feet above the airport elevation.

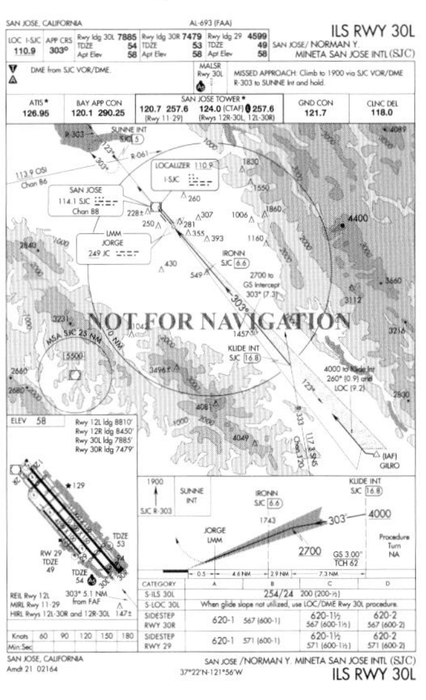

Approximately 240 airports throughout the US currently meet the above criteria.

The initial contour value (lowest elevation) will be at least 500' but no more than 1000' above the airport elevation. The initial contour value may be less than 500' above the airport elevation if needed to depict a rise in terrain close to the runway end. The next contour value depicted will be at a 1000' increment (e.g., 1000'/2000'/3000', etc., NOT 1500'/2500'/3500', etc.). Subsequent contour intervals will be constant and at the most suitable intervals, 1000' or 2000', to adequately depict the rising terrain.

MISSED APPROACH ICONS

Boxed MAP icons, placed in the profile section, are intended to provide quick at-a-glance intuitive guidance to the pilot to supplement, not replace, the textual missed approach instructions in the breifing strip. These step-by-step instructional graphics depict direction of turn, next heading/course/bearing/track, next altitude, etc. to give the pilot the "up and out" initial steps of the missed appraoch.

RNAV CHART MINIMA

RNAV instrument approach procedure charts will now incorporate all types of approaches using Area Navigation systems, both ground based and satellite based. Below is an explanation of the RNAV minima.

The standard format for RNAV minima (and landing minima) is as shown below. RNAV minima are

CATEGORY	A	B	C	D	E
LPV DA	296/40 250 (300 - ¾)				
LNAV/VNAV DA	500/50 454 (500-1)				
LNAV MDA	640/40 594 (600-¾)		640/50 594 (600-1)	640/60 594 (600-1¼)	640-1½ 594 (600-1½)
CIRCLING	640-1½ 594 (600-1½)			640-2 594 (600-2)	740-2½ 694 (700-2½)

dependent on navigational equipment capability, as stated in the applicable AFM or AFMS, or other FAA approved document, and as outlined below.

<u>GLS (Global Navigation Satellite System (GNSS) Landing System)</u>

The GLS (NA) Minima line will be removed from the existing RNAV (GPS) approach charts when LPV minima is published.

<u>LPV (An Approach Procedure with Vertical Guidance (APV) and precise lateral based on WAAS</u>

Must have WAAS (Wide Area Augmentation System) avionics approved for LPV approach.

<u>LNAV/VNAV (Lateral Navigation/Vertical Navigation)</u>

Must have either:

a.) WAAS avionics approved for LNAV/VNAV approach, or
b.) A certified Baro-VNAV system with an IFR approach approved GPS, or
c.) A certified Baro-VNAV system with an IFR approach approved WAAS, or
d.) An approach certified RNP-0.3 system.

Other RNAV approach systems require special approval.

NOTES:

1. LNAV/VNAV minima not applicable for Baro-VNAV equipment if chart is annotated "Baro-VNAV NA" or when below the minimum published temperature, e.g., Baro-VNAV NA below -17° C (2° F).

2. DME/DME based RNP-0.3 systems may be used only when a chart note indicates DME/DME availability; e.g., "DME/DME RNP-0.3 Authorized." Specific DME facilities may be required; e.g., "DME/DME RNP-0.3 Authorized, ABC, XYZ required."

TERMINAL ARRIVAL AREAS (TAAs)

The objective of the Terminal Arrival Area (TAA) is to provide a seamless transition from the enroute structure to the terminal environment for arriving aircraft equipped with Flight Management System (FMS) and/or Global Positioning System (GPS) navigational equipment. The underlying instrument approach procedure is an area navigation (RNAV) procedure. The TAA contains within it a "T" structure that normally provides for a No Procedure Turn (NoPT) for aircraft using the approach. The TAA provides the pilot and air traffic controller with a very efficient method for routing traffic into the terminal environment with little required air traffic control interface, and with minimum altitudes depicted that provide standard obstacle clearance compatible with the instrument procedure associated with it. The TAA will not be found on all RNAV procedures, particularly in areas of heavy concentration of air traffic. When the TAA is published, it replaces the MSA for that approach procedure. TAAs may appear on current and new format GPS and RNAV IAP charts.

The standard TAA consists of three areas defined by the extension of the Initial Approach Fix (IAF) legs and the intermediate segment course. These areas are called the straight-in, left-base, and the right-base areas. TAA area lateral boundaries are identified by magnetic courses TO the IAF. The straight-in area can further be divided into pie-shaped sectors with the boundaries identified by magnetic courses TO the IF/IAF, and many contain stepdown sections defined by arcs based on RNAV distances (DME or ATD) from the IF/IAF. The right/left-base areas can only be subdivided using arcs based on RNAV distances from the IAF's for those areas.

Straight-In Area: The straight-in area is defined by a semi-circle with a 30 NM radius centered on and extending outward from the IF/IAF. The altitude shown within the straight-in area icon provides minimum IFR obstacle clearance

Base Areas: the left and right base areas are bounded by the straight-in TAA and the extension of the intermediate segment course. The base areas are defined by a 30 NM radius centered on the IAF on either side of the IF/IAF. The IF/IAF is shown in the base area icons without its name. The altitude shown within the base area icons provides minimum IFR obstacle clearance.

Minimum MSL altitudes are charted within each of these defined/subdivisions that provide at least 1,000 feet of obstacle clearance, or more as necessary in mountainous areas.

NOTE: Additional information for the TAAs can be found in the Aeronautical Information Manual (AIM) Para 5-4-5-d.

ALTERNATE MINIMUMS

When an alternate airport is required, standard IFR alternate minimums apply. Precision approach procedures require a 600' ceiling and 2 statute miles visibility; nonprecision approaches require an 800' ceiling and 2 statute miles visibility. When a ▲ appears in the Notes section of the approach chart, it indicates non-standard IFR alternate minimums exist for the airport. This information is found in Section E of the TPP. If ▲ NA appears, alternate minimums are not authorized due to unmonitored facility or absence of weather reporting service. Civil pilots see FAR 91.

Take-Off Minimums and (Obstacle) Departure Procedures

When a ▼ appears in the Notes section, it signifies the airport has nonstandard IFR takeoff minimums and/or Departure Procedures published in Section C of the TPP.

CIVIL USERS NOTE: FAR 91 prescribes standard take-off rules and establishes take-off minimums for certain operators as follows: (1) Aircraft having two engines or less - one statute mile. (2) Aircraft having more than two engines - one-half statute mile. These standard minima apply in the absence of any different minima listed in Section C of the TPP.

ALL USERS: Airports that have Departure Procedures (DPs) designed specifically to assist pilots in avoiding obstacles during the climb to the minimum enroute altitude, and/or airports that have civil IFR take-off minimums other than standard, are listed in Section C of the TPP by city. Take-off Minimums and Departure Procedures apply to all runways unless otherwise specified. Altitudes, unless otherwise indicated, are minimum altitudes in MSL.

DPs specifically designed for obstacle avoidance may be described in Section C of the TPP in text or published as a graphic procedure. Its name will be listed, and it can be found in either the TPPs (civil) or a separate Departure Procedure volume (military), as appropriate. Users will recognize graphic obstacle DPs by the word "(OBSTACLE)" included in the procedure title; e.g., TETON TWO (OBSTACLE). If not assigned another DP or radar vector by ATC, this procedure should be flown if visual avoidance of terrain/obstacles cannot be maintained.

Graphic DPs designed by ATC to standardize traffic flows, ensure aircraft separation and enhance capacity are referred to as "Standard Instrument Departures (SIDs)". SIDs also provide obstacle clearance and are published under the appropriate airport section. ATC clearance must be received prior to flying a SID.

NOTE: Graphic Departure Procedures that have been designed primarily to assist Air Traffic Control in providing air traffic separation (as well as providing obstacle clearance) are usually assigned by name in an ATC clearance and are not listed by name in Section C of the TPP.

RNAV Departure Procedures (DP) and Standard Terminal Arrival Routes (STAR)

RNAV DPs and STARs are being developed to support a more efficient traffic flow and further National Airspace System (NAS) capacity. These procedures will be flown only by those aircraft with onboard databases. These procedures will extend over a larger geographic area to allow ATC spacing and sequencing to occur en route. In order to reduce the number of pages required to depict these longer procedures, changes to the graphic depictions and textual data are necessary.

NAVAID boxes will be removed and identified with only the name, the three-letter ident and the applicable symbol. Waypoints will be identified with waypoint symbol and five letter name. Waypoints that overlay NAVAIDs will be depicted only as NAVAIDs, not as a waypoint. A single graphic will be used when possible; however, if not feasible, the common portion of the procedure will be shown on a single page with transitions contained on subsequent pages. Subsequent pages will be subtitled with the transition area, i.e., CHEZZ ONE DEPARTURE Northeast Transitions, or JHAWK TWO ARRIVAL South Transitions. Text remarks that apply to the entire procedure, or all transitions, will be charted on the page that contains the common point and common portion of the procedure. Text remarks that apply to a specific transition will be charted on the page that contains that transition. Transition text will not include a description of the route but will instead state expectations for altitudes, clearances, FL restrictions, aircraft constraints, specific airport arrival use, etc.

There are two types of RNAV SIDs and graphic Obstacle DPs (ODPs): Type A and Type B. Type A generally starts with a heading or vector from the departure runway end and Type B generally starts with an initial RNAV leg near the departure runway end. Type A procedures require the aircraft's track keeping accuracy remain bounded by ± 2 NM for 95% of the total flight time (Type B bounded by ± 1 NM). See the AIM for more specific information.

RNAV Procedures Legs (IAPs, SIDs/DPs and STARs)

Due to the variations in the development, documentation, charting and database coding of RNAV Procedures (IAPs, STARs SIDs/DPs), it has become necessary to chart RNAV legs with specific information based on their type. This data depiction will provide pilots with a clearer indication of the type of leg the aircraft will be flying and the ensuing flight profile.

– Heading - no waypoints shown, "hdg" charted after degrees (i.e., 330° hdg), no mileage shown.
– Direct - waypoint at termination of leg, no course shown, no mileage shown.
– Course - waypoint at termination of leg, course shown, mileage shown only if first leg upon departure.
– Track - waypoints at beginning and termination of leg, course shown, mileage shown.
– Radius - waypoints shown at beginning and termination of leg, no course shown, mileage shown.

Leg mileages will be listed differently based on certain criteria. Mileages on Course and Track legs will be shown to the nearest one-tenth of a NM when all three of the following conditions are met:

Leg termination is 30 NM or less to the Airport Reference Point (ARP) (for STARs, leg origination must be 30 NM or less from the ARP for the primary airport) and,
– leg segment is less than 30 NM and,
– leg segment is not part of the En route structure.

In all other instances, leg mileages will be rounded off to the nearest whole NM, as they are currently.

Instrument Approach Chart Format

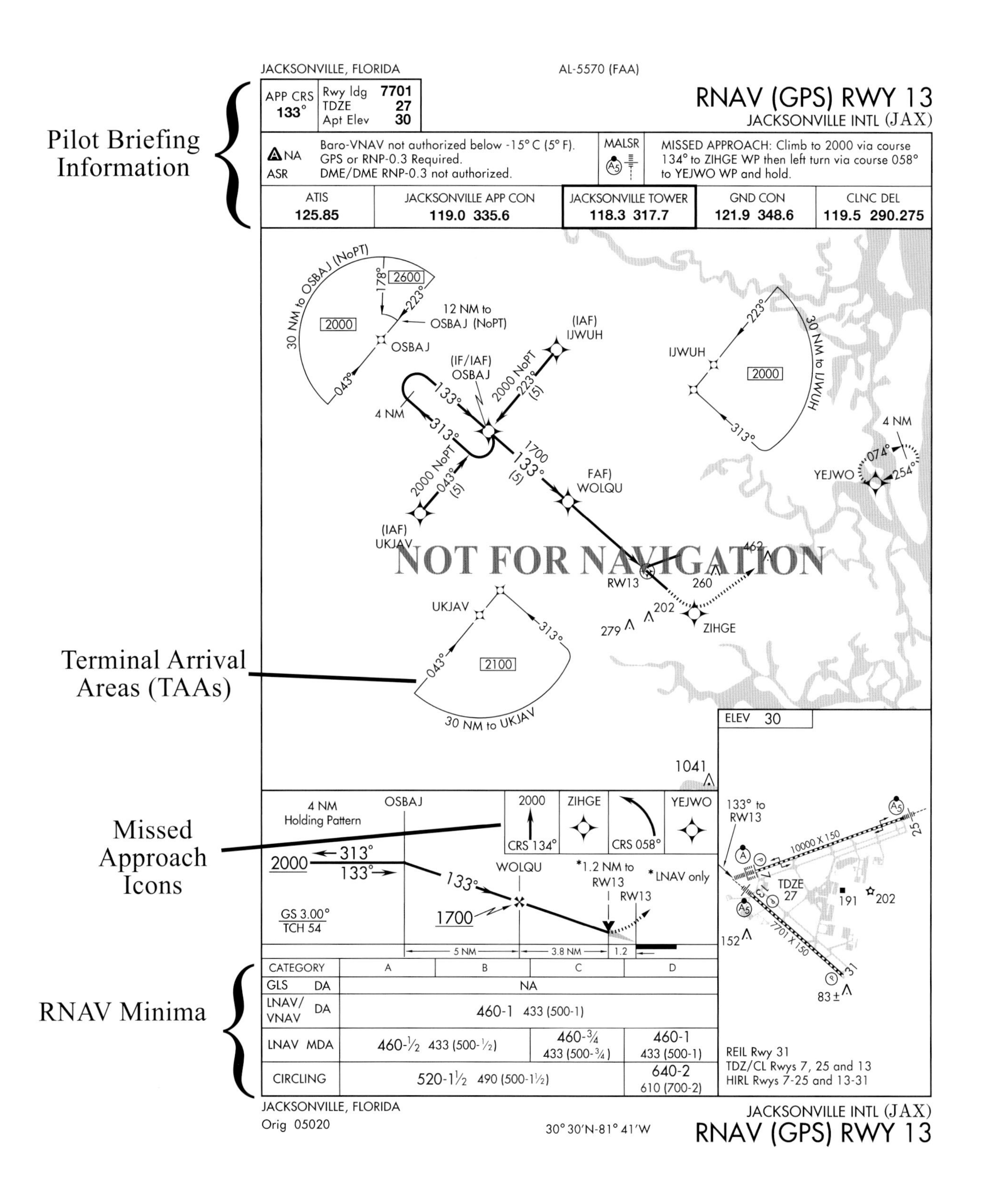

CATEGORY	A	B	C	D
GLS DA	NA			
LNAV/VNAV DA	460-1 433 (500-1)			
LNAV MDA	460-½ 433 (500-½)		460-¾ 433 (500-¾)	460-1 433 (500-1)
CIRCLING	520-1½ 490 (500-1½)			640-2 610 (700-2)

TERMINAL PROCEDURES PUBLICATION SYMBOLS

AERONAUTICAL INFORMATION

GENERAL INFORMATION

Symbols shown are for the Terminal Procedures Publication (TPP) which includes Standard Terminal Arrival Routes (STARs), Departure Procedures (DPs), Instrument Approach Procedures (IAP) and Airport Diagrams.

STANDARD TERMINAL ARRIVAL (STAR) CHARTS
DEPARTURE PROCEDURE (DP) CHARTS

RADIO AIDS TO NAVIGATION

VOR
TACAN
VOR/DME
NDB/DME
VORTAC
LOC/DME
LOC
NDB (Non-directional Beacon)
LMM, LOM (Compass locator)
Marker Beacon
Localizer Course
SDF Course

(T) indicates frequency protection range
Identifier
(Y) TACAN must be placed in "Y" mode to receive distance information
Frequency
ORLANDO
112.25 (T) ORL
Chan 59 (Y)
N28°32.56′ W81°20.10′
Geographic Position
Underline indicates no voice transmitted on this frequency
L-19, H-5
Enroute Chart Reference
DME or TACAN Channel

Coordinates
Waypoint Name
PRAYS
N38° 58.30′ W89°51.50′
Frequency
112.7 CAP 187.1°-56.2
590
Identifier
Reference Facility Elevation
Radial-Distance (Facility to Waypoint)

LOCALIZER 108.5
I-PZV
Chan 22
LOC offset 3.02°
Localizer Offset

REPORTING POINTS/FIXES WAYPOINTS

Reporting Points
N00° 00.00′
W00° 00.00′

75 DME Mileage (when not obvious)

▲ Name (Compulsory)
△ Name (Non-Compulsory)

DME fix

X Mileage Breakdown/ Computer Navigation Fix (CNF)
N00° 00.00′
W00° 00.00′

X (NAME) ("X" omitted when it conflicts with runway pattern)

WAYPOINT (Compulsory)
WAYPOINT (Non-Compulsory)
FLYOVER POINT
MAP WP (Flyover)

STANDARD TERMINAL ARRIVAL (STAR) CHARTS
DEPARTURE PROCEDURE (DP) CHARTS

ROUTES

4500 MEA-Minimum Enroute Altitude
*3500 MOCA-Minimum Obstruction Clearance Altitude
270° Departure Route - Arrival Route
(65) Mileage between Radio Aids, Reporting Points, and Route Breaks
Distance not to scale
Transition Route
R-275 Radial line and value
Lost Communications Track
V12 J80 Airway/Jet Route Identification
(IAS) Holding Pattern
Changeover Point

Holding pattern with max. restricted airspeed
(175K) applies to all altitudes
(210K) applies to altitudes above 6000′ to and including 14000′

SPECIAL USE AIRSPACE

R-352
R-Restricted
W-Warning
P-Prohibited
A-Alert

ALTITUDES

5500 Mandatory Altitude (Cross at)
2300 Minimum Altitude (Cross at or above)
4800 Maximum Altitude (Cross at or below)
2200 Recommended Altitude

MCA (Minimum Crossing Altitude)

Altitude change at other than Radio Aids
All altitudes/elevations are in feet-MSL.
MRA- Minimum Reception Altitude.
MAA- Maximum Authorized Altitude.

AIRPORTS

STAR Charts
Civil
Military
Joint Civil-Military

DP Charts

NOTES

All mileages are nautical.
# Indicates control tower temporarily closed UFN.
★ Indicates a non-continuously operating facility, see A/FD or flight supplement.
All radials, bearings are magnetic.

(NAME2.NAME) - Example of DP flight plan Computer Code.
(NAME.NAME2) - Example of STAR flight plan Computer Code.
SL-0000 (FAA) - Example of a chart reference number.

Alternate Minimums not standard. Civil users refer to tabulation. USA/USN/USAF pilots refer to appropriate regulations.

NA Alternate minimums are Not Authorized due to unmonitored facility or absence of weather reporting service.

Take-off Minimums not standard and/or Departure Procedures are published. Refer to tabulation.

W WAAS VNAV outages may occur daily due to initial system limitations. WAAS VNAV NOTAM service is not provided for this approach.

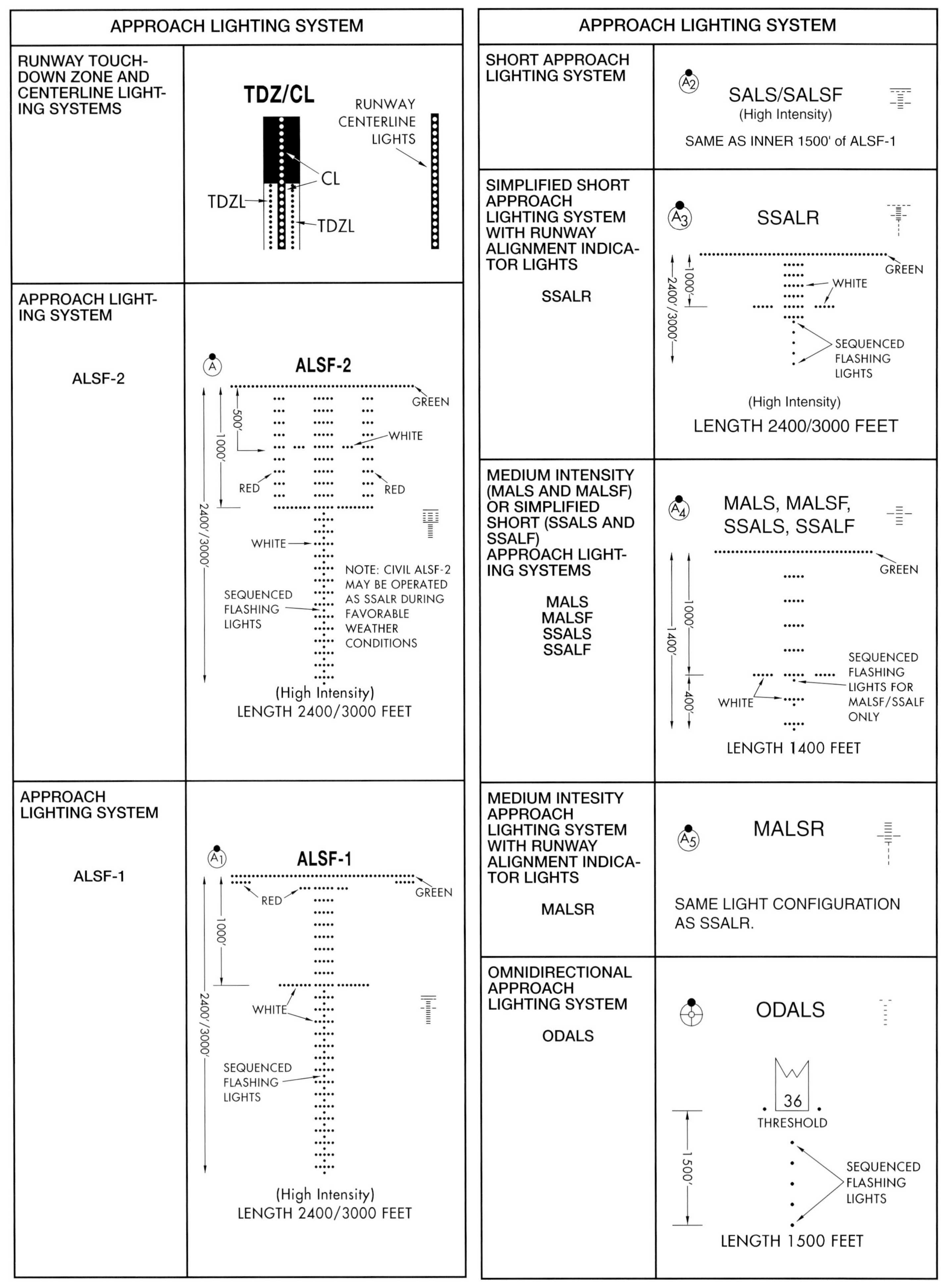
APPROACH LIGHTING SYSTEM
RUNWAY TOUCH-DOWN ZONE AND CENTERLINE LIGHTING SYSTEMS
TDZ/CL
RUNWAY CENTERLINE LIGHTS
CL
TDZL
TDZL
APPROACH LIGHTING SYSTEM
ALSF-2
ALSF-2
GREEN
500'
1000'
WHITE
RED
RED
2400'/3000'
WHITE
NOTE: CIVIL ALSF-2 MAY BE OPERATED AS SSALR DURING FAVORABLE WEATHER CONDITIONS
SEQUENCED FLASHING LIGHTS
(High Intensity)
LENGTH 2400/3000 FEET
APPROACH LIGHTING SYSTEM
ALSF-1
ALSF-1
RED
GREEN
1000'
WHITE
2400'/3000'
SEQUENCED FLASHING LIGHTS
(High Intensity)
LENGTH 2400/3000 FEET
APPROACH LIGHTING SYSTEM
SHORT APPROACH LIGHTING SYSTEM
SALS/SALSF
(High Intensity)
SAME AS INNER 1500' of ALSF-1
SIMPLIFIED SHORT APPROACH LIGHTING SYSTEM WITH RUNWAY ALIGNMENT INDICATOR LIGHTS
SSALR
SSALR
2400'/3000'
1000'
GREEN
WHITE
SEQUENCED FLASHING LIGHTS
(High Intensity)
LENGTH 2400/3000 FEET
MEDIUM INTENSITY (MALS AND MALSF) OR SIMPLIFIED SHORT (SSALS AND SSALF) APPROACH LIGHTING SYSTEMS
MALS
MALSF
SSALS
SSALF
MALS, MALSF, SSALS, SSALF
GREEN
1400'
1000'
400'
WHITE
SEQUENCED FLASHING LIGHTS FOR MALSF/SSALF ONLY
LENGTH 1400 FEET
MEDIUM INTESITY APPROACH LIGHTING SYSTEM WITH RUNWAY ALIGNMENT INDICATOR LIGHTS
MALSR
MALSR
SAME LIGHT CONFIGURATION AS SSALR.
OMNIDIRECTIONAL APPROACH LIGHTING SYSTEM
ODALS
ODALS
36
THRESHOLD
1500'
SEQUENCED FLASHING LIGHTS
LENGTH 1500 FEET

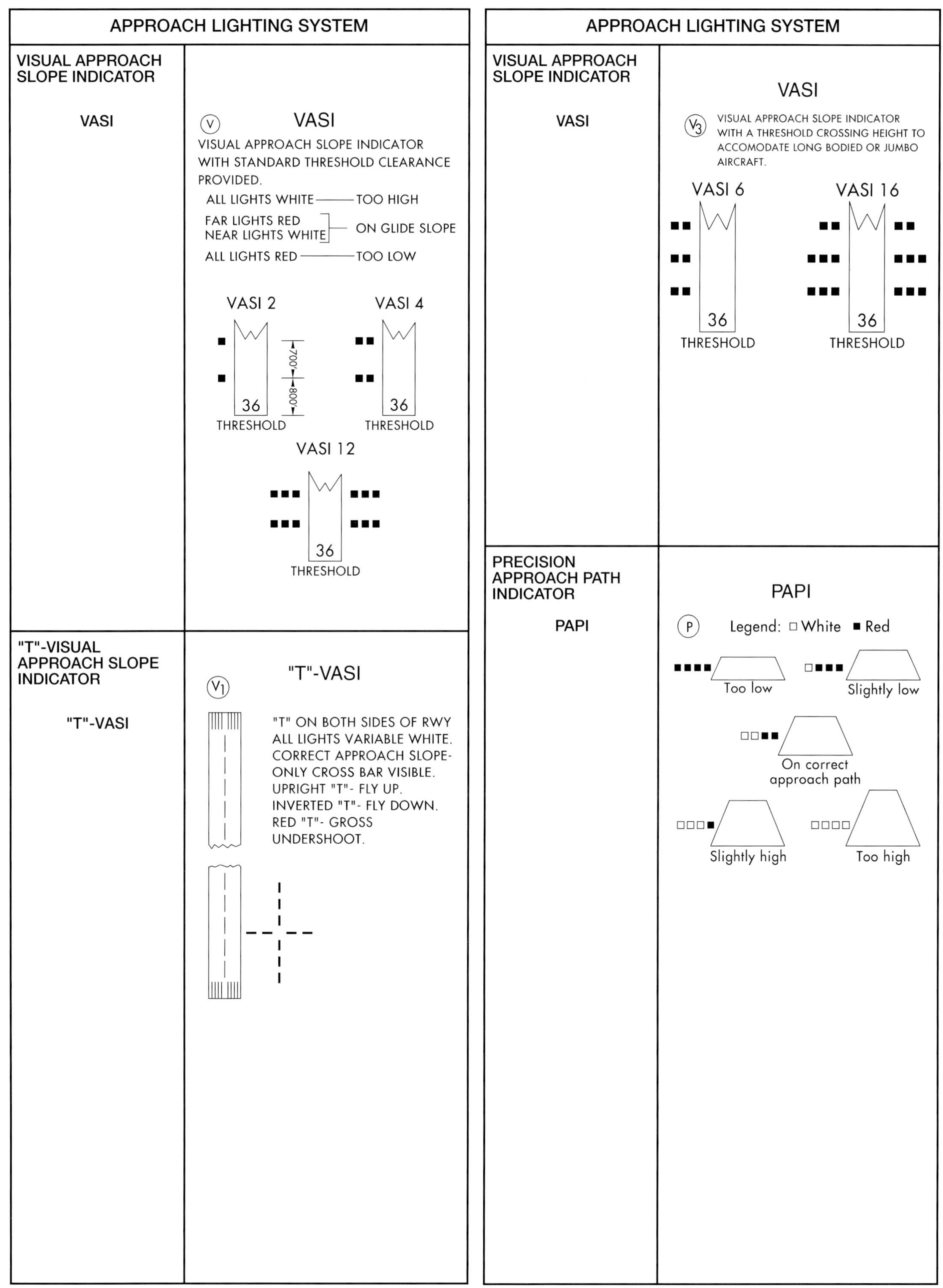
APPROACH LIGHTING SYSTEM
VISUAL APPROACH SLOPE INDICATOR
VASI
V
VASI
VISUAL APPROACH SLOPE INDICATOR WITH STANDARD THRESHOLD CLEARANCE PROVIDED.
ALL LIGHTS WHITE — TOO HIGH
FAR LIGHTS RED
NEAR LIGHTS WHITE — ON GLIDE SLOPE
ALL LIGHTS RED — TOO LOW
VASI 2
700'
800'
36
THRESHOLD
VASI 4
36
THRESHOLD
VASI 12
36
THRESHOLD
"T"-VISUAL APPROACH SLOPE INDICATOR
"T"-VASI
V1
"T"-VASI
"T" ON BOTH SIDES OF RWY
ALL LIGHTS VARIABLE WHITE.
CORRECT APPROACH SLOPE-ONLY CROSS BAR VISIBLE.
UPRIGHT "T"- FLY UP.
INVERTED "T"- FLY DOWN.
RED "T"- GROSS UNDERSHOOT.
APPROACH LIGHTING SYSTEM
VISUAL APPROACH SLOPE INDICATOR
VASI
VASI
V3
VISUAL APPROACH SLOPE INDICATOR WITH A THRESHOLD CROSSING HEIGHT TO ACCOMODATE LONG BODIED OR JUMBO AIRCRAFT.
VASI 6
36
THRESHOLD
VASI 16
36
THRESHOLD
PRECISION APPROACH PATH INDICATOR
PAPI
PAPI
P
Legend: □ White ■ Red
Too low
Slightly low
On correct approach path
Slightly high
Too high

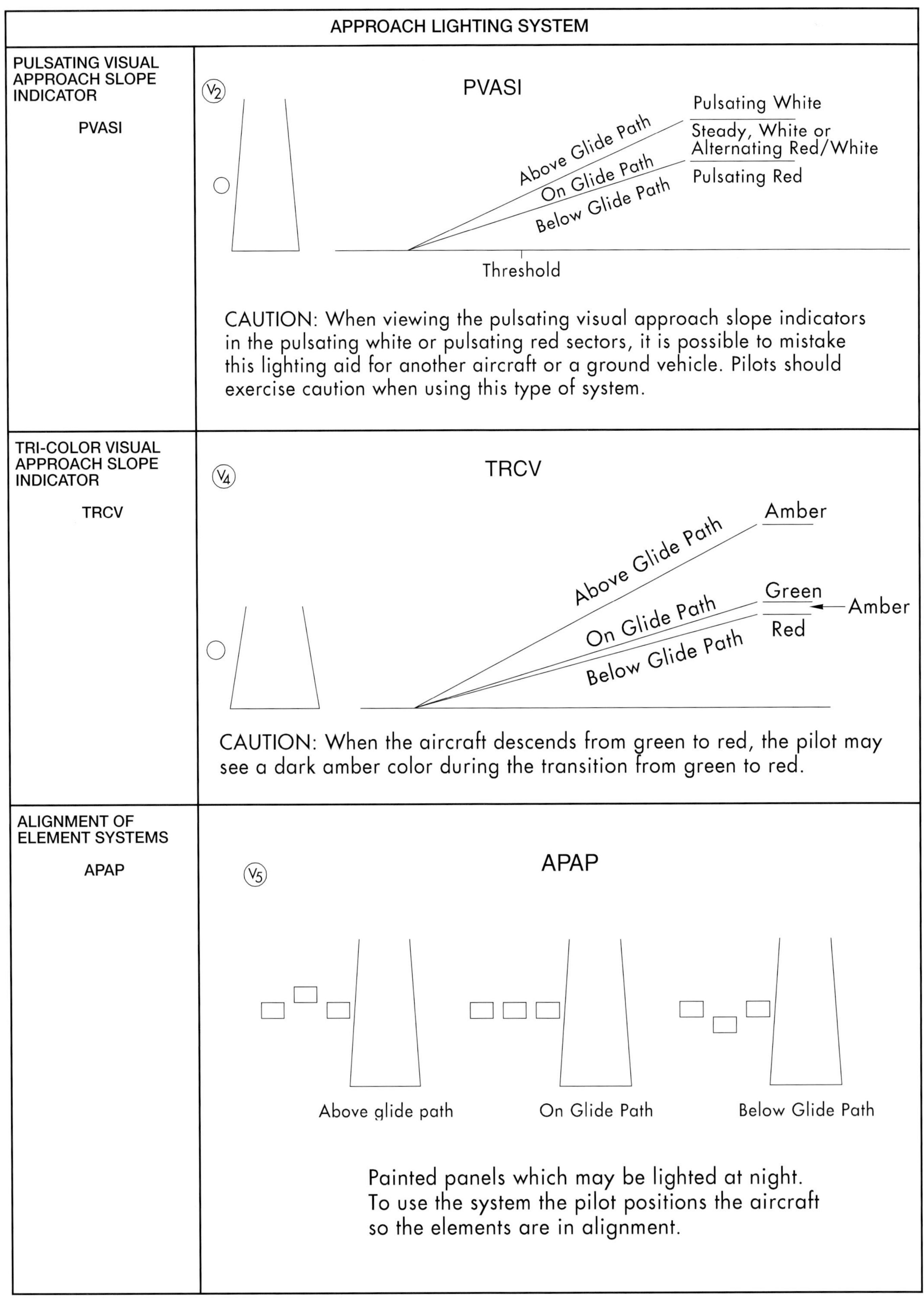
APPROACH LIGHTING SYSTEM
PULSATING VISUAL APPROACH SLOPE INDICATOR
PVASI
V2
PVASI
Pulsating White
Steady, White or Alternating Red/White
Pulsating Red
Above Glide Path
On Glide Path
Below Glide Path
Threshold
CAUTION: When viewing the pulsating visual approach slope indicators in the pulsating white or pulsating red sectors, it is possible to mistake this lighting aid for another aircraft or a ground vehicle. Pilots should exercise caution when using this type of system.
TRI-COLOR VISUAL APPROACH SLOPE INDICATOR
TRCV
V4
TRCV
Amber
Above Glide Path
Green
Amber
On Glide Path
Red
Below Glide Path
CAUTION: When the aircraft descends from green to red, the pilot may see a dark amber color during the transition from green to red.
ALIGNMENT OF ELEMENT SYSTEMS
APAP
V5
APAP
Above glide path
On Glide Path
Below Glide Path
Painted panels which may be lighted at night. To use the system the pilot positions the aircraft so the elements are in alignment.

AIRPORT DIAGRAM/SKETCH

ARRESTING GEAR

- uni-directional
- bi-directional
- Jet Barrier
- Arresting System

ARRESTING GEAR: Specific arresting gear systems; e.g., BAK12, MA-1A etc., shown on airport diagrams, not applicable to Civil Pilots. Military Pilots refer to appropriate DOD publications.

REFERENCE FEATURES

- Buildings
- Tanks
- Obstruction
- Highest Obstruction
- Airport Beacon
- Runway Radar Reflectors
- Hot Spot
- Control Tower #

When Control Tower and Rotating Beacon are co-located, Beacon symbol will be used and further identified as TWR.

Helicopter Alighting Areas

Negative Symbols used to identify Copter Procedures landing point

TDZE 123 Runway TDZ elevation

←0.3% DOWN
0.8% UP→ Runway Slope

(shown when runway slope equals or exceeds 0.3%)

NOTE:
Runway Slope measured to midpoint on runways 8000 feet or longer.

A D symbol is shown to indicate runway declared distance information available, see appropriate A/FD, Alaska or Pacific Supplement for distance information.

AIRPORT DIAGRAM/SKETCH

NOTES

U.S. Navy Optical Landing System (OLS) "OLS" location is shown because of its height of approximately 7 feet and proximity to edge of runway may create an obstruction for some types of aircraft.

Approach light symbols are shown in the Flight Information Handbook.

Airport diagram scales are variable.

True/magnetic North orientation may vary from diagram to diagram

Coordinate values are shown in 1 or ½ minute increments. They are further broken down into 6 second ticks, within each 1 minute increments.

Positional accuracy within ±600 feet unless otherwise noted on the chart.

NOTE:
All new and revised airport diagrams are shown referenced to the World Geodetic System (WGS) (noted on appropriate diagram), and may not be compatible with local coordinates published in FLIP. (Foreign Only)

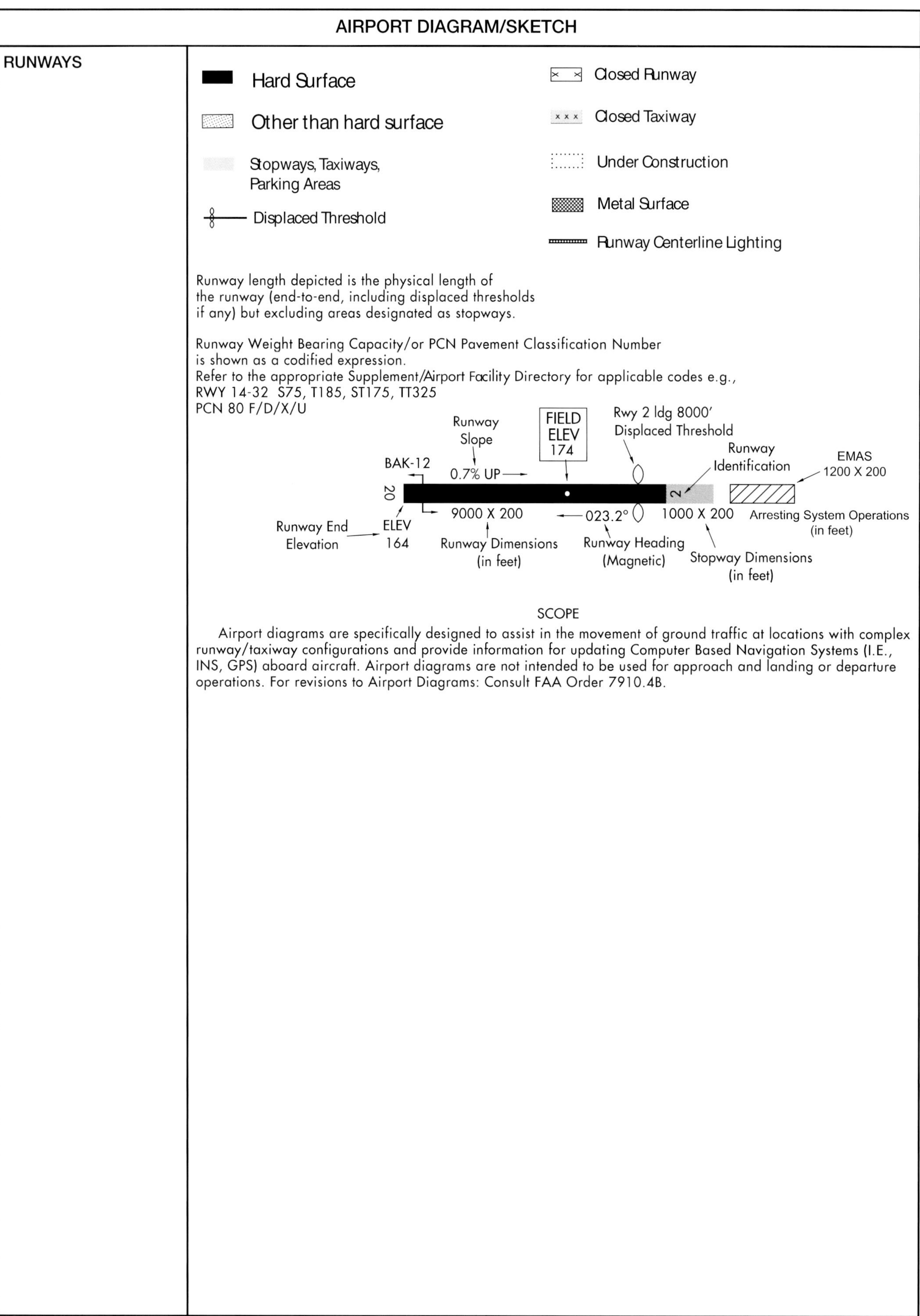
AIRPORT DIAGRAM/SKETCH
RUNWAYS
Hard Surface
Other than hard surface
Stopways, Taxiways, Parking Areas
Displaced Threshold
Closed Runway
Closed Taxiway
Under Construction
Metal Surface
Runway Centerline Lighting
Runway length depicted is the physical length of the runway (end-to-end, including displaced thresholds if any) but excluding areas designated as stopways.
Runway Weight Bearing Capacity/or PCN Pavement Classification Number is shown as a codified expression.
Refer to the appropriate Supplement/Airport Facility Directory for applicable codes e.g.,
RWY 14-32 S75, T185, ST175, TT325
PCN 80 F/D/X/U
Runway Slope
FIELD ELEV 174
Rwy 2 ldg 8000′ Displaced Threshold
Runway Identification
EMAS 1200 X 200
BAK-12
0.7% UP
20
2
9000 X 200
023.2°
1000 X 200
Arresting System Operations (in feet)
Runway End Elevation
ELEV 164
Runway Dimensions (in feet)
Runway Heading (Magnetic)
Stopway Dimensions (in feet)
SCOPE
Airport diagrams are specifically designed to assist in the movement of ground traffic at locations with complex runway/taxiway configurations and provide information for updating Computer Based Navigation Systems (I.E., INS, GPS) aboard aircraft. Airport diagrams are not intended to be used for approach and landing or departure operations. For revisions to Airport Diagrams: Consult FAA Order 7910.4B.

INSTRUMENT APPROACH PROCEDURES PLAN VIEW

TERMINAL ROUTES

Procedure Track

Missed Approached

Visual Flight Path

Procedure Turn (Type degree and point of turn optional)

165°
345°

3100 NoPT 5.6 NM to GS Intcpt
045°
(14.2 to LOM)
Minimum Altitude
2000
155°
Feeder Route
(15.1)
Mileage
Penetrates Special Use Airspace

HOLDING PATTERNS

In lieu of Procedure Turn
270°
(IAS)
090°

Missed Approach
360°
180°

Arrival
HOLD 8000
360°
180°

Limits will only be specified when they deviate from the standard.
Holding pattern with max. restricted airspeed:
(175K) applies to all altitudes.
(210K) applies to altitudes above 6000' to and including 14000'
DME fixes may be shown.
Arrival Holding Pattern altitude restrictions will be indicated when they deviate from the adjacent leg.

REPORTING POINTS / FIXES/ WAYPOINTS

NAVAID Fix

- Compulsory Position Report
- Non-Compulsory Position Report

RNAV Waypoint

- Compulsory Position Report
- Non-Compulsory Position Report

Flyover Point

Intersection

MAP WP (Flyover)

Computer Navigation Fix (CNF)
x (NAME) ("x" omitted when it conflicts with runway pattern)

15 DME Distance From Facility

ARC/DME/RNAV Fix

R-198 Radial line and value

LR-198 Lead Radial

LB-198 Lead Bearing

INSTRUMENT APPROACH PROCEDURES PLAN VIEW

RADIO AIDS TO NAVIGATIONS

VOR

VOR/DME

TACAN

VORTAC

NDB

NDB/DME

LOM/LMM (Compass locator at Outer/Middle Marker)

Marker Beacon

Localizer (LOC/LDA) Course

Right side shading-Front Course; Left side shading-Back Course

SDF Course

180° MLS Approach Azimuth

MLS Identifier
MICROWAVE
Chan 514
M-VDZ
Glidepath 6.20°
DME 111.5 Chan 48(Y)
(Y) TACAN must be in "Y" mode to receive distance information.

LOC/DME

LOC/LDA/SDF/MLS Transmitter (shown when installation is offset from its normal position off the end of the runway.)

LOCALIZER 108.5
I-PZV
Chan 22
LOC offset 3.02°
Localizer Offset

Waypoint Data
Coordinates
Waypoint Name
PRAYS
N38°58.30' W89°51.50'
Frequency
112.7 CAP 187.1°-56.2
590
Identifier
Reference Facility Elevation
Radial-Distance (Facility to Waypoint)

Primary Navaid with Coordinate Values
LIMA
114.5 LIM
Chan 92
S12°00.80'
W77°07.00'

Secondary Navaid
LMM
LIMA
248 NT

MINIMUM SAFE ALTITUDE

Facility Identifier
MSA CRW 25 NM
180°
1500
2200
090°
270°
4500
2500
360°

(arrows on distance circle identify sectors)

INSTRUMENT APPROACH PROCEDURES PLAN VIEW

TERMINAL ARRIVAL AREAS

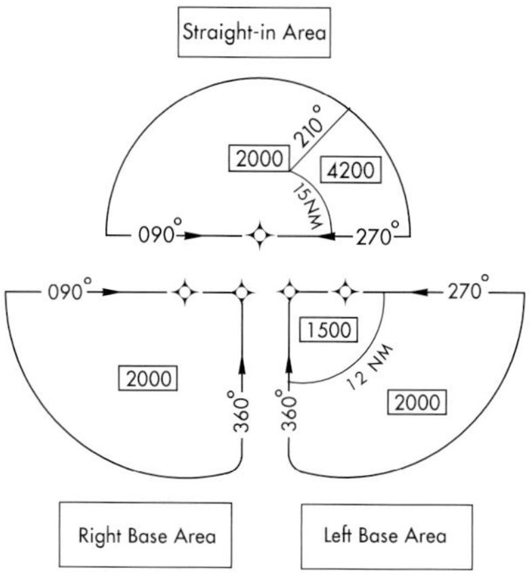

Straight-in Area

2000 4200 210° 15 NM 090° 270°

090° 270° 1500 12 NM 2000 2000 360° 360°

Right Base Area

Left Base Area

Minimum MSL altitudes are charted within each of these defined areas/subdivisions that provide at least 1,000 feet of obstacle clearance, or more as necessary in mountainous areas.

SPECIAL USE AIRSPACE

R-352

R-Restricted W-Warning
P-Prohibited A-Alert

OBSTACLES

- Spot Elevation
- Highest Spot Elevation
- Obstacle
- Highest Obstacle
- ± Doubtful accuracy

FACILITIES / FIXES

FM
IM
MM
NDB
OM
VOR
VORTAC
TACAN
WP

FIX
INT

ALTITUDES

5500 — Mandatory Altitude (Cross at)

2300 — Minimum Altitude (Cross at or above)

4800 — Maximum Altitude (Cross at or below)

2200 — Recommended Altitude

5000 / 3000 — Mandatory Block Altitude

MCA (Minimum Crossing Altitude)

INSTRUMENT APPROACH PROCEDURES PLAN VIEW

MISCELLANEOUS

VOR Changeover Point

RWY 15 S12°00.52′ W77°06.91′ — End of Rwy Coordinates (DOD only)

Distance not to scale

International Boundary

Final Approach Fix (FAF) (for non-precision approaches)

2400 — Glide Slope/Glide Path Intercept Altitude and final approach fix for vertically guided approach procedures.

Visual Descent Point (VDP)

Visual Flight Path

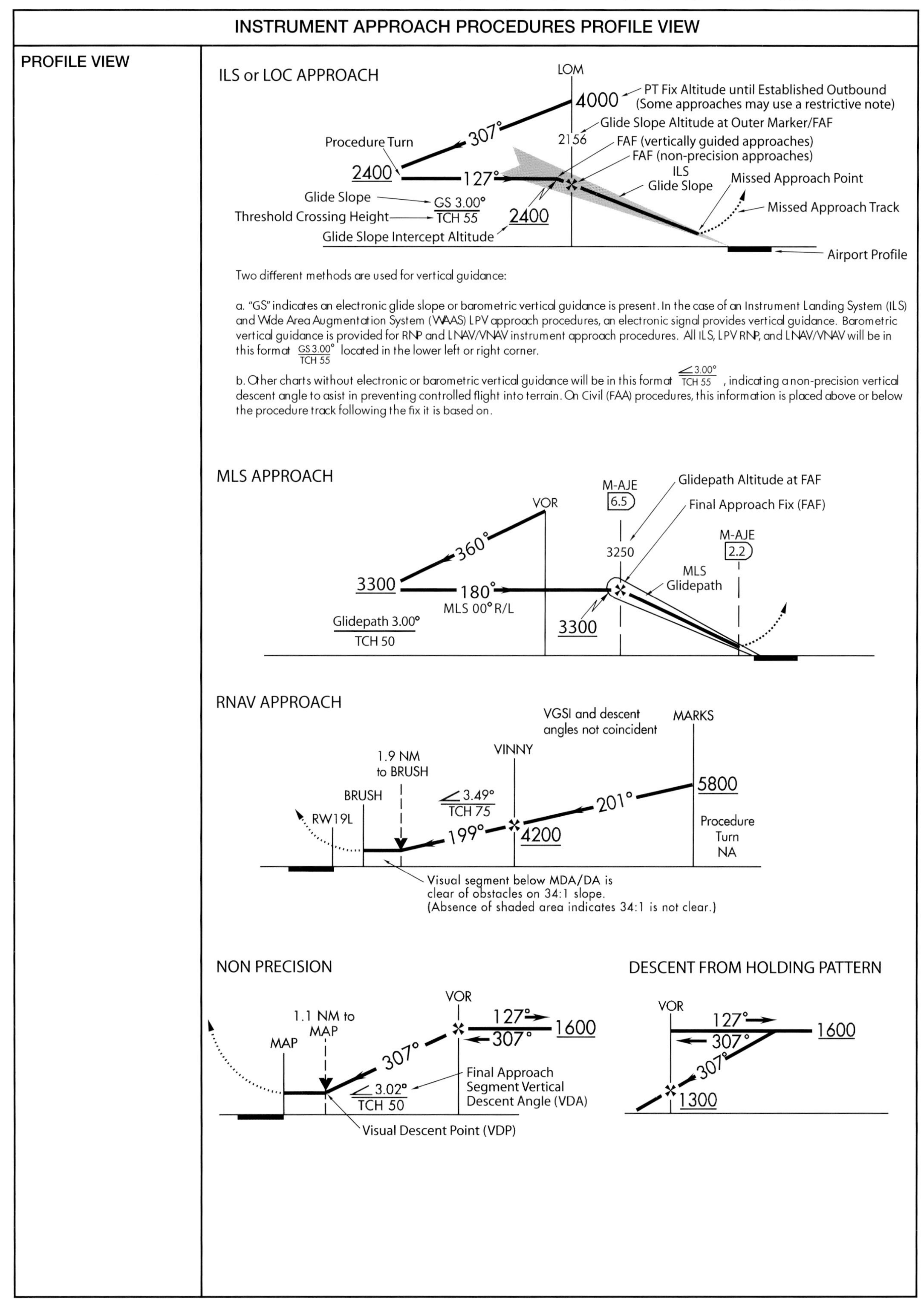

INSTRUMENT APPROACH PROCEDURES PROFILE VIEW

PROFILE VIEW

ILS or LOC APPROACH

Two different methods are used for vertical guidance:

a. "GS" indicates an electronic glide slope or barometric vertical guidance is present. In the case of an Instrument Landing System (ILS) and Wide Area Augmentation System (WAAS) LPV approach procedures, an electronic signal provides vertical guidance. Barometric vertical guidance is provided for RNP and LNAV/VNAV instrument approach procedures. All ILS, LPV RNP, and LNAV/VNAV will be in this format GS 3.00° / TCH 55 located in the lower left or right corner.

b. Other charts without electronic or barometric vertical guidance will be in this format ∠3.00° / TCH 55, indicating a non-precision vertical descent angle to asist in preventing controlled flight into terrain. On Civil (FAA) procedures, this information is placed above or below the procedure track following the fix it is based on.

MLS APPROACH

RNAV APPROACH

NON PRECISION

DESCENT FROM HOLDING PATTERN

ALSO AVAILABLE

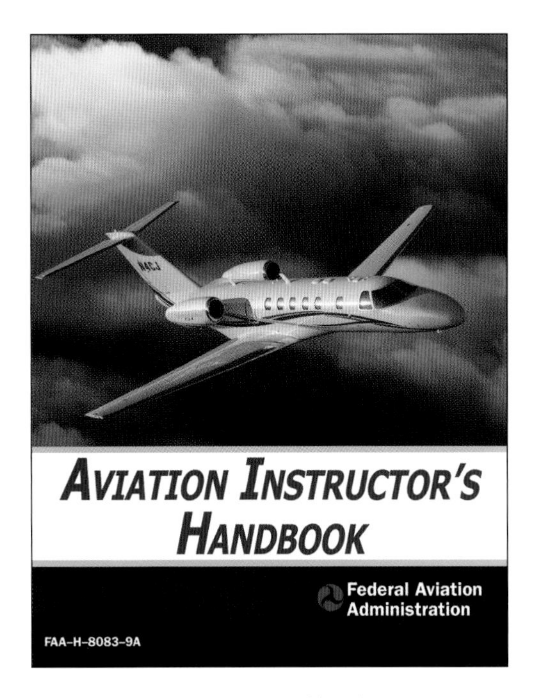

Aviation Instructor's Handbook
FAA-H-8083-9A
Federal Aviation Administration
The official FAA guide—an essential reference for all instructors.
$14.95 Paperback • 228 pages

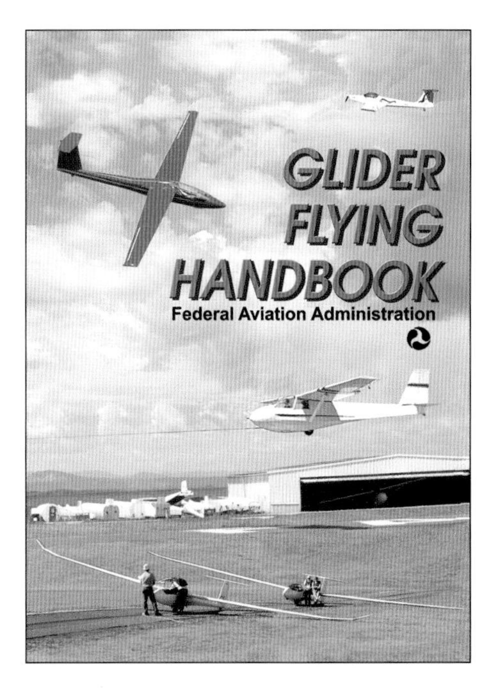

Glider Flying Handbook
Federal Aviation Administration
For certified glider pilots and students attempting certification in the glider category, this is an unparalleled resource.
$24.95 Paperback • 240 pages

Balloon Flying Handbook
FAA-H-8083-11A
Federal Aviation Administration
Essential knowledge necessary for safe piloting at all experience levels. Includes useful illustrations, graphs, and charts.
$16.95 Paperback • 256 pages

Plane Sense
A Beginner's Guide to Owning and Operating Private Aircraft, FAA-H-8083-19A
Federal Aviation Administration
The definitive guide to buying, owning, and maintaining your private aircraft.
$12.95 Paperback • 112 pages

ALSO AVAILABLE

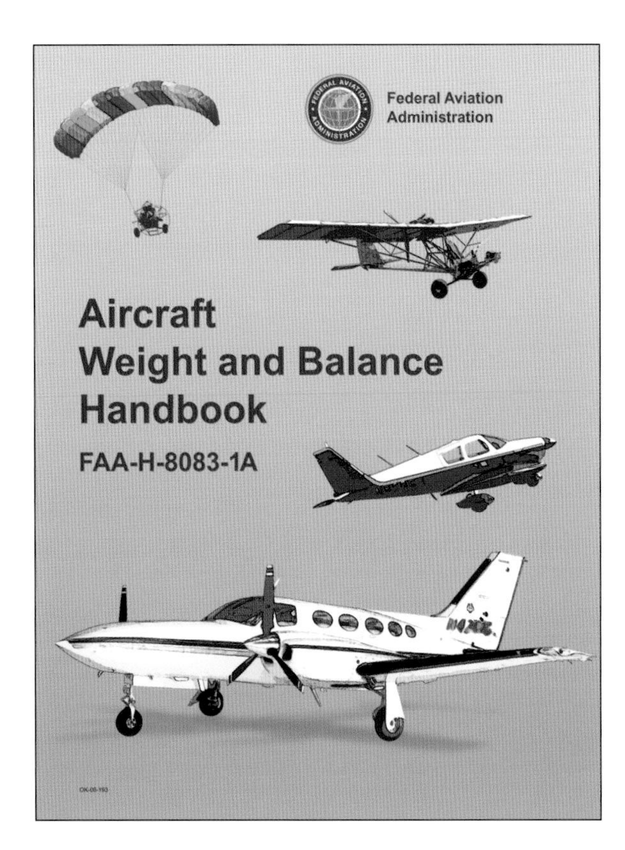

Aircraft Weight and Balance Handbook FAA-H-8083-1A

Federal Aviation Administration

The official FAA guide to aircraft weight and balance.

Aircraft Weight and Balance Handbook is the official U.S. government guidebook for pilots, flight crews, and airplane mechanics. Beginning with the basic principles of aircraft weight and balance control, this manual goes on to cover the procedures for weighing aircraft in exacting detail. It also offers a thorough discussion of the methods used to determine the location of an aircraft's empty weight and center of gravity (CG), including information for an A&P mechanic to determine weight changes caused by repairs or alterations.

With instructions for conducting adverse-loaded CG checks and for determining the amount and location of ballast needed to bring CG within allowable limits, the *Aircraft Weight and Balance Handbook* is essential for anyone who wishes to safely weigh and fly aircraft of all kinds.

$9.95 Paperback • 96 pages

Aviation Weather Services Handbook FAA-AC00-45-F

Federal Aviation Administration and
National Weather Service

A necessary tool for aviators of all skill levels and professions. Includes useful photographs, diagrams, charts, and illustrations.

This official handbook provides an authoritative tool for pilots, flight instructors, and those studying for pilot certification. From both the Federal Aviation Administration and the National Weather Service, this newest edition offers up-to-date information on the interpretation and application of advisories, coded weather reports, forecasts, observed and prognostic weather charts, and radar and satellite imagery. Expanded to 400 pages, this edition features over 200 color and black-and-white photographs, satellite images, diagrams, charts, and other illustrations. With extensive appendixes, forecast charts, aviation website recommendations, and supplementary product information, this book is an exhaustive resource no aviator or aeronautical buff should be without.

$19.95 Paperback • 218 pages

ALSO AVAILABLE

Seaplane, Skiplane and Float/Ski Equipped Helicopter Operations Handbook FAA-H-8083-23

Federal Aviation Administration

The ultimate guide to water-related aircraft piloting.

This comprehensive handbook provides the most up-to-date, definitive information on piloting water-related aircraft. Along with full-color photographs and illustrations, detailed descriptions make complicated tasks easy to understand while the index and glossary provide the perfect references for finding any topic and solving any issue.

The FAA leaves no question unanswered in the most complete book on how to fly water-related aircraft available on the market. The *Seaplane, Skiplane, and Float/Ski Equipped Helicopter Operations Handbook* is the perfect addition to the bookshelf of all aircraft enthusiasts, FAA fans, and novice and experienced pilots alike.

$12.95 Paperback • 96 pages

Powered Parachute Flying Handbook FAA-H-8083-29

Federal Aviation Administration

From the FAA, the only handbook you need to learn to fly a powered parachute.

As far back as the twelfth century, people have loved to parachute. From China's umbrella and Leonardo da Vinci's pyramid-shaped flying device to the first airplane jump in 1912, the urge to leap and soar with the wind has long been a part of history. Parachuting has come a long way since its earliest days due to technological advances, and now more people than ever are taking up this incredible sport. With the *Powered Parachute Flying Handbook* you can make your flying ambitions a reality.

$24.95 Paperback • 160 pages

ALSO AVAILABLE

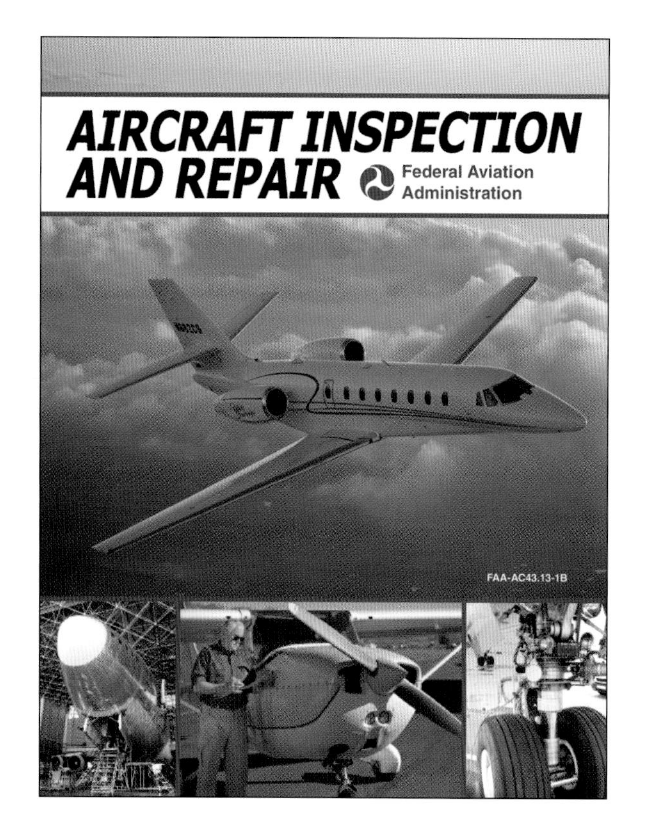

Aircraft Inspection and Repair

Acceptable Methods, Techniques, and Practices
FAA-AC43-13-1B

Federal Aviation Administration

With every deadly airplane disaster or near-miss, it becomes more and more clear that proper inspection and repair of all aircrafts is essential to safety in the air. When no manufacturer repair or maintenance instructions are available, the Federal Aviation Administration deems *Aircraft Inspection and Repair* the one–stop guide to all elements of maintenance: preventive, rebuilding, and alteration. With detailed information on structural inspection, protection, and repair—including aircraft systems, hardware, fuel, engines, and electrical systems—this comprehensive guide is designed to leave no vital question on inspection and repair unanswered. Sections include:

- Wood, fabric, plastic, and metal structures
- Testing of metals and repair procedures
- Welding and brazing, including fire explosion and safety
- Nondestructive inspection (NDI)
- Application of magnetic particles
- Common corrosive elements and corrosion proofing
- Aircraft hardware, from nuts and bolts to washers and pins
- Engines, fuel, exhaust, and propellers
- Aircraft systems and components
- Electrical systems

$24.95 Paperback • 768 pages

ALSO AVAILABLE

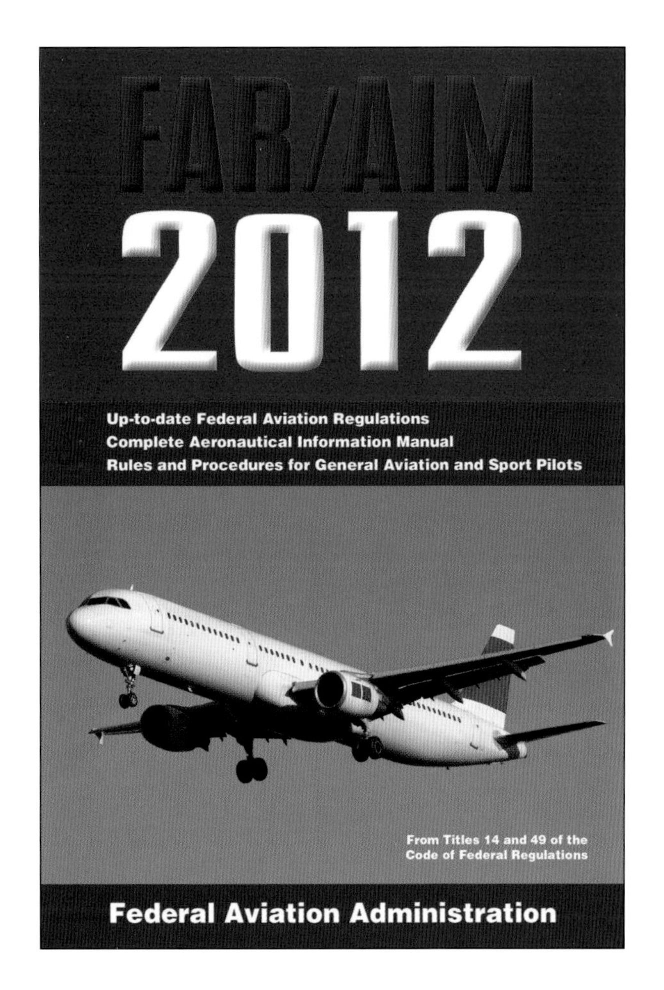

Federal Aviation Regulations / Aeronautical Information Manual 2012

Federal Aviation Administration

As every intelligent aviator knows, the skies have no room for mistakes. Don't be caught with an out-of-date edition of the FAR/AIM. In the current state of aviation regulations, there is simply no excuse for ignorance. In this newest edition of one of the Federal Aviation Administration's most important books, all regulations, procedures, and illustrations are brought up-to-date to reflect current FAA data. With nearly 1,000 pages, this reference book is an indispensable resource for members of the aviation community.

$19.95 Paperback • 960 pages